The International Radio Regulations

Mohamed Ali El-Moghazi · Jason Whalley

The International Radio Regulations

The Case for Reform

Mohamed Ali El-Moghazi
National Telecom Regulatory
Authority of Egypt (NTRA)
Giza, Egypt

Jason Whalley
Newcastle Business School
Northumbria University
Newcastle upon Tyne, UK

ISBN 978-3-030-88570-0 ISBN 978-3-030-88571-7 (eBook)
https://doi.org/10.1007/978-3-030-88571-7

This Springer imprint is published by the registered company Springer Nature Switzerland AG
The registered company address is: Gewerbestrasse 11, 6330 Cham, Switzerland

To Mariam, Laila, and Nour, each of you has a special place in my heart

—Mohamed Ali El-Moghazi

To Treacle, who always found the wrong pile of paper to sit on at the right time

—Jason Whalley

Preface

Every couple of years, spectrum grabs the headlines. It occurred around the turn of the millennium when the auction of 3G spectrum in Germany and the UK raised considerably more than expected, and it is occurring at the moment as campaigners seek to stop the roll out of 5G. For many people, it is the sums raised through the auctions or the health implications of 5G that motivates their interest. They will not have heard of the International Telecommunications Union (ITU), let alone of any of the spectrum related roles that it performs. Yet the ITU is at the heart of spectrum management globally, enabling communication between people and increasing devices to occur. Without the activities undertaken by the ITU, innovations such as mobile phones and Wi-fi, which are taken for granted by many, would not be possible.

Even though the ITU plays a key role in spectrum management, many are unfamiliar with it. They do not know how it is structured or how it operates. And, they are unaware of the radio regulations that govern how spectrum is managed and allocated. This book sets out to change this. It is not, of course, the first book to focus on the ITU. In the 1950s, Codding (1952) wrote the first book to look at the ITU, highlighting the post-WW2 restructuring of the organization that occurred in 1947 and detailing the origin of key concepts such as the international frequency registration board. This was followed in the early 1970s by Levin (1971) who suggested that countries can improve their own spectrum management policies without them being vetoed internationally. The more recent books have examined how the ITU operates, with Manner (2002) outlining the preparations being made for the World Radiocommunications Conference (WRC) within the context of the dot.com crash of the early 2000s. In contrast, Sims, Youell & Womersley (2016) focused on how the ITU-R is participating in forecasting future spectrum needs for IMT.

Although our book builds on these previous efforts, it is also different from them. Previous research typically focuses on the ITU as an international organization rather than on the radio regulations as a major treaty that reflects both international cooperation between countries and technological developments over more than 100 years. Secondly, we highlight the interaction that occurs between national and international spectrum management issues. Of particular importance is the inclusion within the international spectrum management regime of features that are central to what happens nationally. And finally, we consider the impact of recent developments, such as the most recent WRC as well as 5G and cognitive radio systems (CRS), on spectrum management. In other words, we update the discussion of spectrum management within the ITU in the light of the most recent developments in wireless technologies.

Our core argument is that changes are needed to maintain the relevance and usefulness of the ITU. This echoes the call for reform that Michael O'Reilly made in the mid-2010s. While this call for reform is more conciliatory than his threat to defund the ITU, which would, according to O'Reilly, splinter the organization and perhaps cause its functional demise, it highlights the frustration felt by some towards the ITU. But the development of new wireless technologies and services, coupled with changes to the telecommunications industry such as liberalization and deregulation, highlights the continued need for an organization that can regulate spectrum and coordinate the assignments of frequencies to services. The regulation and coordination provided by the ITU facilitates innovation and provides a degree of certainty to the telecommunications industry that encourages them to invest and provide a range of services.

But this is not to suggest that the radio regulations are faultless. One of our aims with this book is to identify areas where changes could be made to ensure the continued relevance of the radio regulations. The telecommunications industry has changed beyond all recognition over the last 40 years due to a combination of liberalization, privatization and innovation. New actors have emerged, which may be international bodies (e.g. IEEE) or companies (e.g. Intel, Microsoft, Facebook), whose ability to participate in discussions around the radio regulations is restricted by the format of World Radiocommunications Conferences. These new actors need to be meaningfully included within discussions around how spectrum is managed globally.

Although structural changes to the ITU have occurred, such as in 1947, the underlying principles of the radio regulations have not changed in over a century. Given all the changes that have occurred, geo-politically and technologically, there is a need to revisit them to ensure their continued appropriateness. In recent years, the role of the ITU-R, which is central to World Radiocommunications Conferences, has come under increasing scrutiny. For around 20 years, decisions within these conferences have been made on the basis of a consensus being achieved—quite simply, everyone needs to agree to whatever is being discussed. The practical implication of this is that the work of the ITU-R over many years on a specific subject can be objected to by a single country. This should encourage discussions between countries as well as the emergence of acceptable compromises. However, there are increasing concerns that the consensus approach is open to abuse, with countries strategically blocking developments to advance their own geo-political interests or to trade support between unrelated issues.

These concerns need to be addressed. Through doing so, we feel, the ITU will be able to continue playing a key role in the global telecommunications industry.

This role provides certainty and legitimacy; enabling countries to manage their spectrum and for mobile operators and device vendors to invest in and develop their businesses. The ITU provides the form for global coordination among stakeholders to occur, stakeholders which are increasing diverse in character, reflecting the far-reaching changes that have occurred over the last 40 years or so. The end result of this coordination is our expanding ability to communicate.

Giza, Egypt — Mohamed Ali El-Moghazi
Newcastle upon Tyne, UK — Jason Whalley
August 2021

References

Codding, G. A. (1952). *The International Telecommunications Union—An experiment in international co-operation.* E.J. Brill: Leiden, The Netherlands.
Levin, H. J. (1971). *The invisible resource; Use and regulation of the radio spectrum.* Johns Hopkins Press: Baltimore.
Manner, J. A. (2002). *Spectrum wars—The policy and technology debate.* Artech House: Boston, Mass.
Sims, M., Youell, T. & Womersley, R. (2016). *Understanding spectrum liberalisation.* CRC Press: Boca Raton, Florida

Acknowledgements

Parts of this book draw on our previous publications. We would like to thank participants at various conferences of the International Telecommunications Society, Telecom Policy Research Conference, Pacific Telecommunications Council and Communication Policy Research South Conference for their comments, criticisms and suggestions over the years. Not only have these helped sharpen our analysis, but the discussions reaffirmed the need to write a book exploring the role of the ITU in spectrum management.

As Chaps. 4 and 5 draw on and extend articles published in Telecommunications Policy, we would like to acknowledge the support of Prof. Erik Bohlin, the editor of this journal, and the reviewers who provided us with helpful and constructive comments. These articles were co-authored with Dr. James Irvine, who not only co-supervized Mohamed during his Ph.D. at the University of Strathclyde (Glasgow, UK), but continued to press him to explain why the ITU is important and how it operates.

We gratefully acknowledge the International Telecommunications Union who granted permission for us to quote from the Radio Regulations, etc., as well as include figures and tables from their publications in the book. Without this, the book would have been very different.

We would also like to thank Prof. Peter Curwen, Mohamed's MCM dissertation supervisor and Jason's frequent collaborator, for the guidance, suggestions and support that he provided while this project was first being discussed.

Finally, while Mohamed is grateful for the support provided by his employer, the National Telecom Regulatory Authority of Egypt (NTRA), he has written this book in a personal capacity. In other words, his opinions should not be interpreted as being those of NTRA.

Contents

Abbreviations

ADS-B	Automatic Dependent Surveillance Broadcast
AI	Agenda Item
AMS	Aeronautical Mobile Service
AMSS	Aeronautical Mobile-Satellite Services
APT	Asia Pacific Telecommunity
ASA	Authorised Shared Access
ASMG	Arab Spectrum Management Group
ATU	African Telecommunications Union
AU	African Union
BEM	Block Edge Mask
BR	ITU-R Bureau
BSS	Broadcasting-Satellite Services
CBRS	Citizens Broadband Radio Service
CCIF	International Long-distance Telephone Consultative Committee
CCIR	International Radio Consultative Committee
CCIT	International Telegraph Consultative Committee
CEPT	European Conference of Postal and Telecommunications Administrations
CITEL	The Inter-American Telecommunications Commission
CPM	Conference Preparatory Meeting
CRS	Cognitive Radio System
DFS	Dynamic Frequency Selection
DSA	Dynamic Spectrum Access
DSL	Digital Subscriber Line
DVB-H	Digital Video Broadcasting – Handheld
EBA	European Broadcasting Area
EC	European Commission
ECC	Electronic Communications Committee
ERMES	Enhanced Radio Messaging System
EU	European Union
FCC	Federal Communications Commission

FCFS	First Come, First Served
FDD	Frequency Division Duplex
FPLMTS	Future Public Land Mobile Telecommunication System
FSS	Fixed Satellite Services
GAA	General Authorised Access
GMDSS	Global Maritime Distress Safety System
GSM	General System for Mobile
GSO	Geostationary Orbit
GSR	Global Symposium for Regulators
HAPS	High Altitude Platform Stations
HDFSS	High-density Applications in the Fixed Satellite Service
IAP	Informatics, Administration and Publications Department
ICAO	International Civil Aviation Organisation
IEEE	Institute of Electrical and Electronic Engineers
IFL	International List of Frequencies
IFRB	International Frequency Registration Board
IMO	International Maritime Organisation
IMT	International Mobile Telecommunications
ITS	Intelligent Transport Systems
ITU	International Telecommunications Union
ITU-D	International Telecommunications Union - Development
ITU-R	International Telecommunications Union - Radiocommunications
ITU-T	International Telecommunications Union - Telecommunications Standardisation
MGWS	Multiple Gigabit Wireless Systems
MIFR	Master International Frequency Register
MSS	Mobile-Satellite Services
NTIA	National Telecommunications and Information Administration
OECD	Organisation for Economic Co-operation and Development
OTT	Over-The-Top
PAL	Priority Access Licensees
PFD	Power Flux Density
PMSE	Program Making and Special Event
PP	Plenipotentiary Conference
PSAA	Procedure for Simultaneous Adoption and Approval
PSD	Power Spectral Density
RA	Radiocommunications Assembly
RAG	Radiocommunication Advisory Group
RCC	Regional Communications in the Field of Communications
RFID	Radio Frequency Identification Devices
RLANS	Radio Local Access Networks
RoP	Rules of Procedures
RR	Radio Regulations
RRB	Radio Regulations Board
RRC	Regional Radiocommunication Conference

RSPG	Radio Spectrum Policy Group
RSTT	Railway Radiocommunication Systems between Track and Train
SAS	Spectrum Access System
SFCG	Space Frequency Coordination Group
SGD	Study Group Department
SMS4DC	Spectrum Management System for Developing Countries
SPTF	Spectrum Policy Task Force
SRD	Short Range Devices
SSD	Space Services Department
TDAG	Telecommunications Development Advisory Group
TDD	Time Division Duplex
TSD	Terrestrial Services Department
TVWS	TV White Spaces
TWIM	Terrestrial Wireless Interaction Multimedia Applications
UAE	United Arab Emirates
UHF	Ultra High Frequency
UN	United Nations
UNDP	United Nations Development Programme
UWB	Ultra-Wide Band
WAPECS	Wireless Access Platforms for Electronic Communication Services
WAS	Wireless Access Systems
WCS	Wireless Communication Service
WIA	Wireless Innovation Alliance
WLAN	Wireless Local Area Network
WMO	World Metrological Organisation
WRC	World Radiocommunication Conference
WTDC	World Telecommunications Development Conference

List of Figures

List of Tables

Chapter 1
Introduction

> *The ITU does not attract the sort of attention and analysis received by other international organizations…The apparent lack of drama at the ITU is one reason…But the importance of telecommunications to modern society makes the ITU too important to ignore.*
>
> James Savage (1989)

1.1 Background

The International Telecommunication Union (ITU) is one of the oldest organizations in the United Nations (UN) system. It predates the UN by more than 75 years, being founded in 1865 to coordinate the international activities of telegraph signals before the emergence of wireless communication. However, James Savage, the author of one of the most important books on the ITU, certainly has a valid point when he noticed how the ITU is usually overlooked in the literature. Needless to say, this is also true for the international radio regulations (RR). Although they do not receive much attention, the RR have had a significant impact on wireless communication, but what are they?

The ITU building in Geneva (Switzerland) contains the Popov Room, named after Professor Alexander Popove, the first person to transmit and receive radio transmissions in 1895. This was just a few years before Marconi sent the first transatlantic radio signal from southwestern England to Newfoundland (Canada) in 1901. These developments ushered in an era of wireless communications, while simultaneously creating issues requiring action.

Radio spectrum was perceived as a new inexhaustible resource, commonly and freely used by the public in a fashion similar to other natural resources. However, two issues came together to suggest the need to manage and regulate spectrum both nationally and internationally. The first of these was interference, which became a major problem as radios transmitted over an increasingly wide band of frequencies and radio emissions crossed frontiers. This caused interference in neighbouring countries. The second issue was the refusal of the Marconi Company to relay messages

M. A. El-Moghazi and J. Whalley, *The International Radio Regulations*,
https://doi.org/10.1007/978-3-030-88571-7_1

received from competing operators. One significant incident occurred when Prince Henry of Prussia attempted to send a courtesy message to the US President as he sailed back home across the Atlantic in 1902. His message was refused by the US shore station because the ship's radio equipment was of a different type and nationality from that onshore.

These two issues, interference and interoperability, encouraged a shift in how the radio spectrum was perceived, from a commodity to a natural resource where usage needs to be managed. The German government convened a conference on wireless telegraphy in Berlin in 1903 with the aim of agreeing on international regulations for radiotelegraph communications. During the conference, two major issues were addressed: the Marconi Company monopoly over radiotelegraphy and the enabled international interconnection.

Shortly afterwards, the first international conference on radiocommunications was held in Berlin in 1906. At the conference, the first International Radiotelegraph Convention was signed between 27 countries, and the first international table of frequencies was agreed. It also established the principle of compulsory intercommunication between land and vessels at sea no matter what radio system was being used. The conference formed the International Radiotelegraph Union with the objective of developing international regulations for spectrum usage. Technical standards for equipment were laid down too, alongside agreements on such matters as the charges for messages.

Perhaps most importantly, from our perspective, was the annex of the 1906 convention that contained the first form of the international radio regulations as we know today. There regulations have been revised numerous times since then, expanding them in length from just a few pages to over a thousand as well as widening their scope so that issues other than interoperability and interference are included.

1.2 The Radio Regulations Simplified

The radio regulations (RR) are one of the instruments of the ITU. They contain a set of international rules regarding the use of spectrum and are ratified by governments and mandatory for all countries. Significantly, the ratification of the ITU Convention by ITU Member States implies acceptance of the RR.

The main principle of the international RR is the sovereign right of each country to assign its frequencies to any service provided that they do not cause harmful interference to other services, which, in turn, operate in accordance with the provisions of the RR. The RR are administered by the ITU-R, the radio sector of the ITU, which is responsible for setting the rules. The RRs, having the status of an international treaty, are binding on all ITU-R Member States.

The international RR are based on two main concepts. Firstly, the allocation of frequency blocks to defined radio services through the Table of Frequency Allocations, as contained in Article 5 of the RR. This concept generally provides common frequency allocations to mutually compatible services operating with similar technical characteristics in specific parts of the spectrum. The second concept is not causing harmful interference to other countries or recognized operating agencies. Historically, the ITU has sought to mitigate harmful interference through dividing spectrum according to the type of service and globally harmonizing its allocation. More specifically, the level of protection required by one type of service may not be commensurate with another type of service.

The ITU-R Table of Frequency Allocation (Article 5 of the RRs) divides the frequency band into smaller bands, which in the 2020 version of the RR are allocated to more than 40 different radiocommunication services. Each spectrum band could be allocated to one or more radio services with equal or different rights (primary and secondary). However, a notable feature is that the providers of a secondary service cannot cause harmful interference to a primary service and cannot claim protection from the harmful interference caused by primary services. The Table of Frequency Allocation contains both primary (printed in 'capitals', e.g. fixed) and secondary (printed in 'normal characters', e.g. mobile) services—see Table 1.1.

The RR divides the world into three regions. Region 1 covers Europe and Africa, while the Americas are to be found in Region 2. The third region covers Asia Pacific and Australia. A frequency allocated in one region can be used in another. For example, frequency band A is allocated to Region 3 but can be reused in Region 1 or 2 for the same or different service.

ITU Member States generally use footnotes to indicate their desire to apply the RR. The footnotes contained in the Table of Frequency Allocations (TFA) can be

Table 1.1 Sample of the RR

<table>
<tr><td colspan="3">415–495 kHz</td></tr>
<tr><td colspan="3">Allocation to services</td></tr>
<tr><td>Region 1</td><td>Region 2</td><td>Region 3</td></tr>
<tr><td>415–435
MARITIME MOBILE 5.79
AERONAUTICAL RADIO NAVIGATION</td><td colspan="2" rowspan="2">415–472
MARITIME MOBILE 5.79
Aeronautical radio navigation 5.77 5.80
5.78 5.82</td></tr>
<tr><td>435–472
MARITIME MOBILE 5.79
Aeronautical radio navigation 5.77
5.82</td></tr>
<tr><td colspan="3">472–479
MARITIME MOBILE 5.79
Amateur 5.80A
Aeronautical radio navigation 5.77 5.80
5.80B 5.82</td></tr>
<tr><td>479–495
MARITIME MOBILE 5.79 5.79A
Aeronautical radio navigation 5.77
5.82</td><td colspan="2">479–495
MARITIME MOBILE 5.79 5.79A
Aeronautical radio navigation 5.77 5.80
5.82</td></tr>
</table>

Source Article 5 of the RR

used in several situations, including to show the status of services (on a primary or secondary basis), as well as whether additional or alternative allocations are to occur.

The RR are revised every three to four years at a World Radiocommunication Conference (WRC). In particular, decisions related to spectrum allocation are taken during ITU-R WRCs. Within WRCs, four main aspects of spectrum management are discussed: the types of service (e.g. fixed and mobile), the types of access (primary and secondary), allocation of geographical areas (global or regional) and technology characteristics (e.g. maximum transmitted power).

Another related document to the RR is the Rules of Procedure (RoPs), which is a regulatory document incorporating the results of meetings of the Radio Regulations Board (RRB). The RRB is composed of 12 elected officials who meet three times per year. The RoPs complement the RR, providing clarification of particular provisions, whose practical application is difficult or may be variously interpreted.

The current RR reflect its evolution since the 1920s. The development of the first frequency allocation table, which dealt with the range of 10 Hz to 60 kHz, was a major achievement of the 1927 conference in Washington. The 1927 Washington DC conference also witnessed the signature of a new Radiotelegraph Convention, the agreement over a new RR and the establishment of the International Radio Consultative Committee (CCIR) where 80 countries would study technical and operating questions related to radiocommunications and to issue recommendations on them. CCIR was renamed ITU Radiocommunication Sector (ITU-R) in 1992.

Another milestones occurred at the 1932 Madrid Conference which expanded the RR to include a table of tolerances and a table giving the acceptable bandwidths for various types of emissions. At that time, the broadcasting and aeronautical services were expanding, and some of the radio stations were outside the international frequency allocation table. This conference also saw the merger between the International Telegraph Conference and the International Radiotelegraph Conference to form the ITU.

At the International Telecommunication Convention of 1932, the term telecommunication was defined for the first time by the ITU as follows: '*Any telegraphic or telephonic communication of signs, signals, writing, facsimiles and sounds of any kind, by wire, wireless or other systems or processes of electric signaling or visual signaling (semaphores)*'. In addition, the word radiotelegraph was replaced by radiocommunication, reflecting the evolution of technology. It was defined as: '*Any telecommunication by means of Hertzian waves*'.

In 1947, the International Radio Conference held in Atlantic City (USA) decided that all users must abide by the TFA. The conference also created the International Frequency Registration Board (IFRB) to regulate the use of spectrum.

In 1982, the format of the frequency band was changed—an upper and lower frequencies were provided, thereby defining the band. In 1992, the Additional Plenipotentiary Conference in Geneva undertook a reform of the ITU to provide it with greater flexibility so that it could adapt to an increasingly complex, interactive and competitive telecommunications environment. The ITU's three main areas of activity were organized in 'Sectors': telecommunication development, radiocommunications and telecommunication standardization. As part of this reform, the CCIR was renamed Radiocommunication Sector (ITU-R).

1.3 The Purpose of This Book

The RR are a well-established treaty that have been accepted and adopted globally. But when they were established over a century ago, they reflected a set of principles that may not be valid today. While the telecommunication industry has witnessed significant changes since the privatization and liberalization movements began in the mid-1980s and wireless technologies have been through numerous advancements, the main principles underpinning the international spectrum management regime have remained unaltered.

Countries have sought to manage spectrum by designating appropriate uses, technologies and users via a regime of common and control (C&C). Such an approach has been heavily criticized for creating an artificial scarcity that is due to inefficient utilization rather than spectrum shortage. Since the 1990s, an increasing number of issues, many of which are related to one another, have emerged that collectively question the wisdom of continuing with an unaltered set of principles guiding how spectrum is allocated and managed. As a spectrum shortage has emerged due to high demand for data services, an intense debate has emerged regarding whether the traditional C&C approach should be retained.

Although alternatives to C&C have been suggested, their application has not occurred save for in specific contexts such as 3G mobile auctions and technology neutrality within the International Mobile Telecommunications (IMT) standards, or for particular technologies such as Wi-Fi in the ISM bands. As a result, in most of the world and especially in developing countries, radio spectrum continues to be managed traditionally.

This book aims to investigate one potential factor that is largely overlooked in the debate, that is, the interaction between national policies and the international spectrum management regime administrated by ITU-R and underpinned by the international RRs. Through exploring the interplay between national and international spectrum management, we shed light on the relationship between the two and how they influence one another. By doing so, the possibility of the C&C approach being replaced, either in its entirety or partially, by one of three alternatives (spectrum market, commons and easements) is investigated.

Needless to say, its importance lies in the fact that the international spectrum management regime was created in parallel with the formulation process of national policies as technologies were invented and wireless communications became commonplace to handle interference and to enable international interoperability. The main principles underpinning the regime have remained more or less constant, which is arguably surprising given the widespread changes that have occurred and the regular opportunities for revision, as demonstrated through WRCs and other conferences, which exist. In other words, the global spectrum management regime could have been changed, but for whatever reason it has not.

The right context to change the current spectrum policy would appear to be the RRs, which are already accepted by 193 countries. Therefore, irrespective of whether the international spectrum management regime is a restriction on the overhaul of national policies, it has a large influence depending on the country, and it is a reflection of what most of the stakeholders in the spectrum policy domain require. That is why, in this book, the reader will be exposed to the different elements of the international spectrum management regime and its associated RRs that shape spectrum policy nationally.

This book also sets out to find a meeting ground between the ground-breaking revolutionary ideas on spectrum policy that have been discussed in the literature in recent decades and have been partially implemented in a few developed countries and the legacy RRs validated by the ITU-R and pursued by most of the world. To this end, we engage in a novel analysis of the international spectrum management regime has been conducted. The book addresses the emerging involvement and increasing influence of developing countries in the ITU-R and undertakes an in-depth analysis of the International Mobile Telecommunication (IMT) standardization process and the a priori planning for broadcasting services.

1.4 Structure of the Book

The rest of this book is structured as follows. Chapters two and three present the necessary literature needed to understand the principles of spectrum management at both the national and international levels. Chapter two provides background on radio spectrum and its management. It analyses national radio spectrum management policies and covers the four main approaches to spectrum management nationally: C&C, spectrum market, spectrum commons and spectrum easements.

Chapter three analyses the main principles of spectrum management at the international level. It starts with a brief overview of the ITU-R as an international organization and international regime and traces the origins of the international spectrum management regime. The chapter analyses the international regimes in terms of the same four elements that were used in analysing the national spectrum management policies.

Chapter four discusses radiocommunication service allocation. It starts with the perceptions and applicability of the allocation nationally. The chapter then explores the interaction between the international spectrum management regime and national spectrum management policies with regard to radiocommunication service allocation flexibility. The bulk of the chapter, however, focuses on the main elements of the international service allocation framework.

Chapter five addresses technology selection. It starts with recounting how technology selection is perceived and discusses its application. The chapter then explores the interaction between the international regime and national policies with regard to decisions related to technology neutrality and technology selection. Attention then turns to the influence of IMT standardization on the definition of mobile technology generations and on the discrimination between generations of technology. The chapter also explores the influence of IMT spectrum identification on technology selection and technology neutrality and the influence of IMT standardization on the decline of WiMAX.

Chapter six focuses on the opportunistic access concept. It starts with an overview of the concept and its applicability in practice before exploring the interaction between the international regime and national policies with regard to its adoption. The chapter draws on empirical data to understand perceptions of opportunistic access, radiocommunication service status and the influence of WRC-12 decisions of Agenda Item 1.19 on Cognitive Radio System (CRS).

Chapter seven tackles assignment rights, the fourth dimension of spectrum management. It starts with a discussion of the different models of spectrum assignment before focusing on spectrum refarming and spectrum pricing with a discussion of national practice. Finally, the chapter addresses the impact of the RRs on spectrum assignments nationally. In Chapter eight, the focus is on developing countries. In particular, we focus on the process of radio spectrum policy reform in developing countries. The chapter also addresses the changing presence of developing countries in the ITU-R in recent years, focusing on WRC-12 and the additional mobile allocation in the 700 MHz band to illustrate the changes that have occurred.

Chapter nine presents a regime theory analysis of the international spectrum management regime in terms of the main actors, norms and principles. It also challenges the traditional view that the ITU-R is purely a technical rather than a political body. The chapter seeks to ascertain the dominant regime theory that applies to modern international spectrum management.

The subject of chapter ten is the most recent WRC, which was held in Egypt in 2019. The chapter illustrates the decision-making processes that occur and shape the international spectrum management regime, using IMT as the vehicle to demonstrate the changing role of developing countries at the conference as well as how decisions are the result of a series of trade-offs between stakeholders. The chapter also demonstrates how WRC are part of a larger process, composed of committees, conferences and working groups that collectively shape the international spectrum management regime.

The final chapter reflects on the challenges to the international spectrum management regime that have discussed over the course of the book. These challenges, which have varied in their severity, have challenged the 'status quo' of the international spectrum management but they have not toppled it. The resilience of the regime is a testament to its acceptance, with the consequence that the suggested ways of changing it that we discuss occur within the context of its global continuation.

Chapter 2
National Radio Spectrum Policy

Advocates of a unique solution to the exclusive vs. shared approach dilemma have been wasting everybody's time by promoting the virtues of a single-sided approach to reforming radio wave management'.

Freyens (2010)

2.1 Introduction

In recent years, many countries have reviewed their spectrum policies. Countries have moved away from the traditional administrative approach, famously called 'command and control', towards an approach that is mainly driven by market forces, by technological innovation or a combination of both. The reform process was largely motivated by the liberalization and privatization of the telecommunication sector. While many aspects of the sector have changed, governments have often acted cautiously when it comes to spectrum, preferring not to assign spectrum solely through one approach in preference to another. Some spectrum has been reserved for unlicensed operations, while elsewhere it has been licensed.

Before the rest of this book focuses on the international radio regulations (RR), this chapter addresses spectrum policy at the national level. With this in mind, this chapter shall start with a brief introduction of radio spectrum as a natural resource before moving onto the function of spectrum management. This is then followed by an analysis of national radio spectrum management policy along four dimensions—radiocommunication service allocation, technology selection, spectrum usage rights and frequency assignments. Section 2.4 then explores the different approaches to spectrum management adopted nationally—command and control, spectrum market, spectrum commons and spectrum easements—before Sect. 2.5 evaluates these alternatives from three distinct approaches (technical, economic and regulatory). Conclusions are drawn in Sect. 2.6.

M. A. El-Moghazi and J. Whalley, *The International Radio Regulations*,
https://doi.org/10.1007/978-3-030-88571-7_2

2.2 Radio Spectrum

The word spectrum means '*the distribution of a characteristic of a physical system or phenomenon or the range of values of a quantity or set of related quantities*' (Dictionaries, 2001). Radio is defined by the ITU as '*Electromagnetic waves of frequencies arbitrarily lower than 3000 GHz, propagated in space without artificial guide*' (ITU-R, 2020). Radio spectrum is the distribution of the radio frequencies, and each frequency represents a portion of the radio spectrum. The term 'spectrum' is usually used instead of 'radio spectrum' to indicate the same meaning.

Spectrum is a finite but non-exhaustible resource. It is finite because the range of radio frequencies that are suitable for wireless communications is limited to spectrum bands from 9 kHz to 3000 GHz. It is non-exhaustible because it is infinitely renewable and is not consumed by use. Moreover, spectrum is a non-homogenous resource as it has different characteristics according to the frequency band. For instance, the reusing of spectrum below 100 MHz is difficult because it is capable of propagating over very long distances. In contrast, spectrum above 5 GHz propagates over short distances (Cave et al., 2007). These propagation characteristics limit what use the spectrum can be put to. The use of the same frequencies by multiple users at the same place and time leads to usage distortion for all or some of the users—this effect is called 'interference'. To address interference, some form of coordination between those sharing the band is needed. Spectrum can be shared across four dimensions: frequency, spatial location, time and signal separation (ITU-R, 2001a).

Radio spectrum is integral to delivering wireless services. TV, radio, GPS and Wi-Fi are all examples of services that spectrum can offer to the public. Spectrum was first utilized for wireless transmission in 1895 when Alexander Popov and Guglielmo Marconi transmitted a wireless signal over a short distance in, respectively, Russia and Italy. Shortly afterwards, the human voice was first broadcast in 1906 (Timofeev, 2006), and, in the same year, one of the first commercial radio systems was built by the Marconi Wireless Telegraph Company that linked together coastal radio stations and ships (Anker & Lemstra, 2011).

2.3 Radio Spectrum Management

In general, spectrum management is the allocation of different spectrum bands to different radiocommunication services (e.g. fixed, mobile), authorizing users to access particular parts of the spectrum, managing type approval and electromagnetic compatibility standards for wireless devices and monitoring spectrum use to restrain unauthorized uses (Cave et al., 2006). However, the way spectrum is managed has evolved over time. In particular, during the early day of wireless communications, spectrum was perceived as a new inexhaustible natural resource commonly and freely used by the public (Struzak, 2003). However, the issue of interference became apparent for a number of reasons. Firstly, radios transmitted over a wide band of frequencies due to the lack of knowledge regarding how to tune transmitters and receivers. Secondly, wavelength measures could not be estimated before 1905. Thirdly, many operators preferred to transmit over a wide range of frequencies to avoid the need to precisely adjust receivers. Finally, transmitters used a lot of power to reach long distances. All of these issues called for changing the view of spectrum, away from a commodity towards spectrum being a scarce resource that needs to be managed (Kruse, 2002).

In 1925, the US Secretary of Commerce declared that there was no space in the spectrum for further assignments and thus stopped issuing new licenses. A year later, the US Attorney General declared that the secretary had no authority to define any rights to spectrum. Surprisingly, instead of appealing against such a decision, the secretary decided to issue open access licenses without any fees. This led to a chaotic situation that was not resolved until the Federal Radio Act of 1927. The act envisioned spectrum as a public property that is regulated by the Federal Radiocommission, which later became the Federal Communications Commission (FCC). In the UK, wireless communications were regulated by the Telegraph Acts of 1868 and 1869. In 1904, the Post Office was considered as the wireless regulator, and the government nationalized the private wireless shore stations (Marcus, 2004).

These developments ushered in a period of strict spectrum regulation by governments for several reasons (Horvitz, 2013). The first was national security, while the second was the immaturity of radio technology in its early days. As clarified by Horvitz (2013: 8): *'Government regulation was another way to compensate for the hardware's inadequacy'*. The third reason was the Marconi monopoly over wireless telegraphy, which motivated governments to establish their national sovereignty over radio spectrum. Freyens (2009) points out that the governmental approach was the conventional way to manage spectrum from the early 1930s until 1994 when the FCC conducted its first spectrum auction (Cramton, 2002).

2.4 An Analysis of National Radio Spectrum Management Policies

National spectrum management policies vary around the world according to local regulations and circumstances. For instance, in the USA, the National Telecommunications and Information Administration (NTIA) is responsible of assigning frequencies to all federal government owned and operated radio stations (Cave & Morris, 2005). In contrast, in Tunisia, there is an entity separate from the regulator that is responsible for spectrum management (Agence Nationale des Fréquences (ANF), 2012). Therefore, it is not an easy task to identify common principles or approaches. There have, however, been some attempts to analyse national spectrum management policies in terms of their main components.

Chaduc and Pogorel (2008) undertake a review of the different radio spectrum management approaches and suggest that there are four elements that determine the type of spectrum management approach or policy: service allocation, technology selection, usage rights and assignment type. Moreover, Freyens (2009) adopts a slightly different approach to the categorization, using usage flexibility, rights exclusivity, club membership, rules control and rights assignment instead. Here we will adopt the four element categorization system of Chaduc and Pogorel (2008), largely due to its clarity and simplicity.

2.4.1 Service Allocation

The first element, service allocation, is the distribution of spectrum to the different radiocommunication services (Foster et al., 2011) where a service is defined as *'transmission, emission and/or reception of radio waves for specific telecommunication purposes'* (ITU-R, 2020). More specifically, it determines the use of a given frequency band by one or more terrestrial, space radiocommunication or radio astronomy services (ITU-R, 2008). The service allocation can be flexible/neutral or harmonized. Harmonization refers to the common designation of bands for particular radiocommunication services in different countries (Indepen & Aegis Systems, 2004), whereas service flexibility/neutrality implies that any radiocommunication service can be offered in the frequency band (Frullone, 2007).

Harmonization is useful in mitigating interference, reducing cross-border coordination requirements, promoting international mobility and decreasing equipment costs. However, harmonization could also lead to restrictions on the use of under- or unused spectrum for alternative uses and on the ability to refarm spectrum for new services (Chaduc & Pogorel, 2008; Indepen & Aegis Systems, 2004). Therefore, Cave (2002) argues that harmonization should be time limited until it enables manufacturers and operators to deliver a cost-effective service. After that, the market should be opened to other services.

It should also be noted that, in theory, there are other ways to allocate spectrum other than by service (Eurostrategies & LS-Telecom, 2007; ITU-R, 2001b). For instance, spectrum could be allocated based on grouping services together that have similar characteristics together or by categorizing services according to the service area of the radio system application (e.g. terrestrial point-to-point, space earth-to-space). Spectrum could also be divided into bands where low, medium and high-power services exist, or divided on the basis of the amount of interference one could expect to encounter when using a specific band of spectrum.

2.4.2 Technology Selection

The second element of spectrum management policy is technology selection. The selection of technologies could be neutral, restricted to standardized technologies or selective of specific technologies. Firstly, standardization refers to the level of specification of allocated services such as transmitter power, channelization and interoperability. On the other hand, technology neutrality is defined by Foster (2008) as the minimum applied constrains while ensuring that interference is appropriately addressed. Whittaker (2002) argues that true technology neutrality implies defining conditions without any biased assumptions. Furthermore, an important element of technology selection is channel planning, which is defined by Chaduc and Pogorel (2008) as the intermediate stage between harmonization of service allocation and technology standardization. Channel planning accommodates type of channel duplex mode—frequency division duplex (FDD) and time division duplex (TDD)—and the width of the channel (e.g. 10, 20 MHz).

The main advantage of standardization is that it allows large-scale production that reduces the cost of equipment. Moreover, the benefits of standardization include avoiding harmful interference and promoting interoperability between terminals and public networks. Anker and Lemstra (2011) argue that national regulators usually prefer to select a standardized technology because it obviates the need for specifying particular technology characteristics as part of the licensing process. The license conditions would instead have to state that the operator's technology should conform to a global standard. Having said this, standardization may lead to lock-in to an inferior standard and delay the introduction of new equipment (Indepen & Aegis Systems, 2004; London Economics, 2008; Pogorel, 2007). An example of the failure of regulators to select the right technology is enhanced radio messaging system (ERMES), which was an initiative to create a European-wide mobile messaging system that ended with no significant implementation (Cave, 2002).

2.4.3 Spectrum Usage Rights

The third element of spectrum management policy is usage rights. Chaduc and Pogorel (2008) define three categories of spectrum usage rights: exclusive property rights, exclusive property rights with easement and collective (non-exclusive).

Exclusive property rights in spectrum entail the licensing of clean spectrum, free of interference, to the licensee. While this encourages operators to take financial risks and invest in establishing wireless networks, it may create entry barriers and decrease competition (Chaduc & Pogorel, 2008). On the other hand, easement is a certain right to use the property of another without possessing it. Hence, exclusive spectrum property rights with easements allow other users to access spectrum that is owned or licensed to a particular entity without causing harmful interference. Easements exist in many cases, but to different degrees. For instance, some spectrum bands accommodate sharing between fixed terrestrial and satellite radiocommunication services with users of the two services having exclusive property rights (Chaduc & Pogorel, 2008). The third type is collective usage rights, which involves common access to the spectrum without having exclusive property rights in the spectrum by a particular entity. While this has the benefits of decreasing entry barriers and promoting innovation, it adds significant technical constraints and increases the risk of interference (Chaduc & Pogorel, 2008).

There are differences between the type of usage rights and the type of spectrum access. Exclusive property rights means that spectrum is exclusively owned by an entity that has the right to determine access rules to the spectrum. This access could be individual or collective. To elaborate, let us consider three different cases. The first case represents a company that has a license to use a single frequency in the VHF band for corporate communications. The second case is where a mobile operator has an exclusive wide frequency band license to provide services to end users. The third case is a common frequency band that is open for usage by the public (e.g. 2.4 GHz). While in the first case of access to spectrum is for individuals, in the second and third cases, access is on a collective basis. Furthermore, the first and second cases represent exclusive property rights in terms of their usage rights, while the third case entails collective usage rights.

A question that consequentially emerges is what types of usage individual and collective access to the spectrum are associated with? To answer this question, it is useful to consider Netting's observation of the communal and private forms of land tenure in a Swiss village. Netting (1976) found that both forms coexisted where each is suitable for a particular type of land use. Netting (1976) also noted that communal forms of land tenure are mostly associated with a low value of production per unit area, low frequency and dependability of use or yield, low possibility of improvement or intensification, large area required for effective use and large labour and capital-investing groups. Outside of these areas, private forms of land tenure are used.

Mapping this analogy onto spectrum, it can be argued that collective access to spectrum is a better approach whenever the value of a single frequency is small. Therefore, access to a wide range of frequencies is needed to reap the maximum benefits from the spectrum. In addition, if users are less dependent on the spectrum and have low frequency of use, there would be no incentive for them to have an exclusive spectrum license. Moreover, if production based on spectrum as an input can be improved, this encourages users to invest in individual licenses. Furthermore, if a required spectrum band for a particular application is quite large, wide non-exclusive spectrum is arguably more suitable. Finally, an individual license will be impractical if benefits from spectrum require large labour and investment. An example of this is DSL and leased line. Both are used to provide access to the Internet and data services; however, the latter is used whenever access to the Internet is more frequent and the business depends on data services. Another example is GSM, provided by large operators that made it impractical for any individual to invest in a private network to provide a similar (voice) service. The following table summarizes the difference between individual and collective access to the spectrum (Table 2.1).

Table 2.1 Individual and collective spectrum access

Nature of spectrum usage	Spectrum access type	
	Individual	Collective
Value of single frequency	High	Low
Frequency and dependability of use	High	Low
Possibility of production improvement	High	Low
Size of required spectrum	Small	Large
Required investment and labour	Small	Large

2.4.4 Frequency Assignment

The fourth element of spectrum management policy is frequency assignment, where there are three categories: administrative, market based or license exempt (Chaduc & Pogorel, 2008). Administrative assignments could be conducted through a first-come first-served basis, beauty contest, comparative bidding or a lottery (ERC, 1998). In the former, the license applications are dealt with in the order of their receipt, and the license is granted when the applicant fulfils the stated criteria. In contrast, in a beauty contest, applicants submit their proposals that are then assessed, while in a comparative bidding process applicants submit a cash bid in addition to their proposals. In lotteries, regulators choose randomly from the applicants.

There are two types of assignments within the market-based approach: spectrum auction and spectrum trading (secondary market). Auctions are a transparent and flexible means of applying price mechanisms to spectrum (Cave & Valletti, 2000), whereas spectrum trading is a way of applying the free market to spectrum assignments that can be transferred to other users as long as there is no harmful interference (ITU-R, 2004). It is worth mentioning that both trading and auctions are not mutually exclusive. Spectrum could initially be auctioned and then traded. Finally, in license exempt assignments, access to spectrum does not require a license from the regulator.

2.5 National Radio Spectrum Management Policy Approaches

The combination of the four aforementioned elements results in several different approaches to spectrum management. However, there are no common view on the number and naming of the different national spectrum management approaches. For instance, Pogorel (2007) has the view that there are nine national spectrum management policies. He refers to the 'open access' approach proposed by Noam (1995) as 'California Dream' and the 'easements' approach proposed by Faulhaber and Farber (2003) as 'Technical command and control plus Mitigated Market'. The next four subsections will concentrate on those approaches that are the most common in the literature.

2.5.1 Command and Control

The command and control approach is based on allocation harmonization, standardized technology, exclusive property rights and administrative assignment (Chaduc & Pogorel, 2008). Within such an approach, the regulator manages spectrum by designating appropriate uses, technologies and users (OECD, 2006). This traditional approach to spectrum management worked well for many years for a number of reasons. Firstly, there was a small number of wireless services that evolved at a slow pace (Wellenius & Neto, 2005), with sufficient frequencies being available to accommodate most of the users (Wellenius & Neto, 2005). Thirdly, the command and control approach was, and still is, useful with regard to achieving international spectrum harmonization for a particular service and facilitating the adoption of global technical standards to provide interoperability and economies of scale (OECD, 2006). It is also a key tool available to national regulators to achieve public interest goals such as coverage requirements and ubiquity of services (Cave et al., 2007).

In addition, controlling the use of spectrum makes it easier to manage harmful interference as regulators would be able to model the interaction between neighbouring services and tailor license conditions accordingly (Cave et al., 2007). Furthermore, command and control, it has been argued, is suitable for services such as broadcasting where governments tend to control the number and identity of broadcasters (Cave, 2006). Finally, the command and control approach is aligned with the view that telecommunications industry is a 'natural monopoly' that is not suited to competition (Ratto-Nielsen, 2006).

Several developments have called for a review and reform of the command and control approach. The first issue is the liberalization of telecommunications markets, which started in the USA and spread around the world that resulted in a competitive wireless industry (Ratto-Nielsen, 2006). Shortly afterwards, the FCC in the USA was the first regulator to adopt an auction in spectrum assignment in 1994 (Kelly & LaFrance, 2012). The second issue was the rapid growth in wireless services, especially mobile phone services, which increased the demand for spectrum (Wellenius & Neto, 2005). Most of the valuable spectrum had already been assigned in many countries to particular users or entities (Wellenius & Neto, 2005). It was, therefore, difficult for new services to find vacant spectrum that could accommodate their needs.

Thirdly, operators had also been facing a shortage in the spectrum needed to meet the growing broadband demand. In particular, the growth in mobile data traffic was almost doubling every year (Cisco, 2012). In addition, Cisco forecast that global mobile data traffic would increase 18-fold between 2011 and 2016 (Cisco, 2012). The introduction of 4G services has generated more data traffic—a 4G connection generates 28 times more traffic, on average, than a non-4G connection (Cisco, 2012). It is argued that the reason behind such growth in data is the increasing utilization of video services such as movies on demand and the emergence of widely used Internet social networks such as Facebook. Furthermore, operators need spectrum not only to address capacity issues but also to provide new applications and services as well as to compete with over-the-top (OTT) service providers.

Although adopting more efficient technologies is expected to provide an increase of between three and 10 times mobile data capacity by 2030 (Ofcom, 2012), the issue is that while peak wireless spectral efficiency is doubling every 30 months, user demand for bandwidth doubles at a much faster rate (Bennett, 2010). In addition, the advances in mobile broadband technologies have enabled them to be close to the Shannon limit of a wireless channel's data rate. Moreover, achieving the high data rates of the new technologies require large carrier bandwidth of up to 100 MHz, which is not available in most of the spectrum bands allocated to 3G and 4G services.

Fourthly, several technical studies have indicated that spectrum utilization efficiency is quite low. For instance, a study by shared spectrum showed that 62% of the 30 MHz–3 GHz spectrum band is unutilized during peak hours in a dense urban area (The New America Foundation and The Shared Spectrum Company, 2003). Another study showed that 95% of the government's spectrum in the USA is not being used at any given time (Economist, 2004). Technological innovation has called for a review of the way that spectrum is managed. Technologies such as spread spectrum and cognitive radio enable more efficient use of the spectrum and decrease the possibility of interference (Wellenius & Neto, 2005).

In addition to all of the previous issues, there are several other problems associated with the command and control approach. For instance, such an approach could create an artificial scarcity due to spectrum underutilization and inefficiency rather than an actual shortage in the available spectrum (Wellenius & Neto, 2005). Moreover, Lehr (2005) points out how this artificial scarcity distorts the opportunity cost of spectrum. More specifically, the opportunity cost is high for new users who desire to adopt new technologies and services because of the high spectrum fees and restricted access to spectrum. In contrast, the opportunity cost is low for governmental agencies and incumbents as they face little incentive to invest in enhancing their spectrum usage by adopting new technologies or other means.

Cave et al. (2006) explain that regulators traditionally have focused on promoting the public interest while delaying the introduction of new technologies and services and artificially increasing service costs. Hazlett (2001) asserts that the 'public interest' concept is vague, which gives the regulator a maximum degree of freedom to restrict access to spectrum. He further explains that regulators usually tend to monitor inefficiencies resulting from overutilization rather than those associated with underutilization. Furthermore, Marcus (2004) argues that the history of radio proves that governments have suppressed more technologies than they have promoted. For instance, although spread spectrum technology was invented in the 1940s, it remained classified by the government until it was reinvented in the 1960s.

Taken together, these issues have motivated several regulators to revise their spectrum policies. For instance, in the USA, the NTIA reviewed its fundamental spectrum policy objectives in 1989 (FCC, 2002). Following this review, the Spectrum Policy Task Force (SPTF) was established in 2002, with one of its recommendations being that spectrum policy should be a balance between the granting of exclusive spectrum usage rights through market-based mechanisms and spectrum commons on the one hand and using command and control in limited circumstances on the other (FCC, 2002). In Europe, the European Commission issued a green paper in 1998 on radio spectrum policy requesting comments on issues such as the harmonization of radio spectrum allocation and radio spectrum assignment and licensing (European Commission, 1998). And, in the UK, Professor Martin Cave was appointed in 2001 to lead an independent review of radio spectrum management (Cave, 2002). Cave (2002) called for a market-based approach that enabled spectrum trading in the commercial use of spectrum and gradually extended the market mechanism to other public services.

2.5.2 *Spectrum Market*

The spectrum market approach is based on allocation flexibility, technology neutrality, exclusive property rights and market-based assignment (Chaduc & Pogorel, 2008). In such an approach, it is argued, private operators are more knowledgeable of the market and will choose the optimum service for consumers (Hazlett, 2001). The origin of the market approach can be traced back to the dilemma that the FCC faced in choosing between three different standards for colour television signal transmitting systems in 1951 (Herzel, 1951). The FCC performed what can be called a technical beauty contest and adopted only one standard. Herzel (1951) suggests a revolutionary solution at that time would have been to use an auction to choose among the competing standards. Moreover, Herzel (1998) explained more recently that his main concern at that time was not using an auction as much as having packages of spectrum rights and obligations. Smythe (1952) defended the FCC position by explaining that different spectrum users should be treated differently according to their activities, and, therefore, bidding cannot be used as the mechanism to choose among them. Herzel (1952) responded by stating that radio spectrum is similar to any other object (e.g. equipment) that users compete for.

Coase (1959) supported Herzel's argument in his seminal contribution 'The Federal Communications Commission' and suggested that spectrum assignments should be treated in a similar fashion to property rights. He argued that the aim of the regulator should be to maximize the output and not to minimize the interference, and that interference should not be an issue as long as the gain from it is more than the harm it produces. Interference could be resolved by limiting the user's spectrum usage rights and could even be accepted by users if they were paid more than the value by which their service was decreased.

Coase subsequently generalized his argument in his Nobel Prize article 'The Problem of Social Cost', suggesting an approach towards dealing with harmful effects on others that is based on comparing the total social product yielded by alternative arrangements (Coase, 1960). Therefore, interference inflicted by one user on another could be allowed if societal benefits are much more rather than in the case of limiting interference. Rothbard (1982) opposes this view and argues that the first user of a resource has the property rights so that the last user should be responsible of resolving the interference. By applying Coase's view to the issue of interference in wireless communication, Cave and Webb (2003) suggest that interference could be allowed into neighbouring bands if the value of the increase in capacity and/or reduction in equipment cost is greater than the value of the loss in the neighbouring side due to the interference.

According to Baumol and Robyn (2006), there are three main elements of the market-based approach. The first is to design exclusive property rights for the use of the spectrum in a specific geographic area and applying technical rules to limit the interference between licensees. The second is using the auction method for the initial allocation of spectrum rights and applying a secondary market method. The third is allowing maximum flexibility in the types of commercial services that can be provided in the licensed spectrum. In the following subsections, each of these is examined in turn.

2.5.2.1 Auction

Spectrum auctions were firstly introduced in the USA in 1994 and since then there have been calls to expand the FCC's auction authority to encompass new frequency bands and new wireless services including unlicensed services (Kelly & LaFrance, 2012). The FCC was also the first to use a simultaneous multiple round auction design that allows participants to bid for spectrum in different territories at the same time (Oranje et al., 2008). The success of spectrum auctions has encouraged Kwerel and Williams (2002) to suggest that the FCC should auction spectrum voluntarily offered by incumbents and any unassigned spectrum in a large-scale two-sided band restructuring auctions, in what was described as a 'Big Bang Auction'.

Arguments in favour of auctions include them being a transparent flexible means of employing price mechanism to the spectrum while achieving policy objectives (Cave & Valletti, 2000). In addition, auction bids reflect the price of anticipated services, and, therefore, higher auction fees do not imply higher prices for the end-users (Hazlett, 1998). On the other hand, auctions are argued to create barriers to small entrants and reduce the number of free and non-profitable services. Moreover, actions are used to raise revenue for governments (Noam, 1998).

Auctions have been widely used around the world for cellular communication since the FCC auction in 1994, raising more than $100 billion by 2010 (McAfee et al., 2010). They have even been utilized to allocate satellite services in countries such as Brazil (Nordicity, 2010). However, Cave and Foster (2010) explain that the practice of spectrum auctions has not proved successful in some cases for a variety of reasons: political interference (preferences and reserve prices), the failure to perceive the importance of the interaction of auctions among different countries, high opening bid prices set by the regulator, inadequate spectrum packages, inappropriate reserve prices, unrealistic expectations by bidders, flawed bidding rules, collusion by participants and excessive service obligations. There are also concerns that auctions usually result in raising the value of a license and in delaying the deployment of services (Cave, 2002).

2.5.2.2 Trading

It has been argued that spectrum markets enhance the utilization efficiency of spectrum provided that property rights are fully and precisely defined, and there are no transaction costs (Cave & Webb, 2003). However, Cave (2008) points out that although auctions are useful in initially allocating the spectrum to those who have the highest value for it, the trading of spectrum between the different users ensures that this is the case even if the valuation of spectrum changes over time. One argument in favour of spectrum trading is that it creates incentives for users to apply their spectrum to the highest-valued uses as determined by the market (OECD, 2006) and that it also removes those artificial scarcities arising from the administrative allocation of the spectrum (Wellenius & Neto, 2005). Spectrum trading also offers auction bidders a safety net in case their business proves to be unsuccessful and could also provide companies that lost out in an auction with the opportunity to acquire a license, possibly at a reduced price (WIK, 2006).

There are, of course, some concerns—trading might increase interference among users and constrain harmonizing services in adjacent spectrum bands (OECD, 2006). Perhaps the main problem associated with trading is that users do not know whether they will be affected by interference until the transmission begins (Cave & Webb, 2012). In addition, trading might also encourage spectrum hoarding by incumbents seeking to restrict competition (Wellenius & Neto, 2005). Furthermore, trading cannot be adopted in cases where current spectrum fees do not reflect spectrum economic value (Falch & Tadayoni, 2004). Finally, trading could lead to extensive fragmentation of spectrum assignment where extensive negotiations between users and many guard bands are needed (Indepen, 2001).

In general, the practice of trading has not proven to be that successful in either the USA or UK (Akalu & Arias, 2012; Benkler, 2011; Cave et al., 2007). It is argued that the nature of spectrum services limits the effectiveness of trading as there are generally three types of spectrum licenses (El-Moghazi, Whalley, & Irvine, 2012). The first one includes a large number of licensees with a small band of spectrum (such as PMR). In such cases, the trading reflects changes in ownership and is relatively limited in number. For instance, the percentage of PMR trading transactions in the UK, a pioneering country when it comes to spectrum trading, comparing to the total number of licenses, is just 4.3% (Electronic Communications Committee, 2011).

In the second type, there is small number of licensees, such as mobile operators, who have a large band of spectrum. Here the market is relatively stable, and spectrum is a valued asset with the consequence that the number of trading transactions is almost none. In the third type of license, spectrum is customized to the licensee's business. For instance, a fixed link license is used to connect two sites of the licensee. Such a license is critical to the licensee and of no importance to others. In France, where there are 30,000 fixed link licenses and trading is allowed, no transaction has occurred (Electronic Communications Committee, 2011). In addition, the nature of wireless services—the dominance of cellular mobile services, uncertainty surrounding interference from incompatible technologies, etc.—limits the effectiveness of spectrum trading or spectrum property rights.

2.5.2.3 Flexibility

Spectrum flexibility entails that the holder of spectrum property rights has the ability to determine the uses that the spectrum can be put to. The practice of introducing flexibility varies across the world. One example of flexibility was when the FCC auctioned spectrum for multipurpose 'wireless communications service' in 1997. However, the auction resulted in little revenue due to the uncertainty regarding the uses that the spectrum could be put to (Goodman, 2004). In Europe, a concept called 'Wireless Access Platform For Electronic Communication Services' (WAPECS) was introduced to enable more flexibility within the EU spectrum management framework through allowing the use of spectrum on a technology and service neutral basis within certain limits. The initiative was based on replacing the traditional service and technology combination by what was called 'electronic communications services', where mobile, portable or fixed access services could be provided under one or more frequency allocations (mobile, broadcasting, fixed), deployed via terrestrial and/or satellite platforms using a variety of technologies. The industry response to WAPECS was that they preferred harmonized frequency arrangements and that the cost associated with a more flexible use of spectrum in terms of band plan neutrality was too high (CEPT, 2010).

Furthermore, in El Salvador, a liberal approach to spectrum management occurred in 1997. Spectrum assignments were technologically neutral, although the regulator allocated specific service to each spectrum band, users can deviate from such allocation without a penalty. In addition, assignments cover 20 years and are auctioned in case of multiple applicants and assignments can be transferred or subdivided in terms of their frequency, geographical coverage and time dimensions without regulatory approval (Hazlett & Muñoz, 2006).

2.5.3 Spectrum Commons

Before examining the spectrum commons approach, it is useful to firstly explore the commons concept where it was firstly used in medieval times when woodland and pasture were set aside for the joint use of villagers (Vogler, 2000). In addition, Benkler (2003) perceived commons as a particular type of institutional arrangement for governing the use and disposition of resources whether they could be opened to anyone or only to a defined group and whether it is regulated or not. Chaduc and Pogorel (2008) define a spectrum commons as an approach that is based on service flexibility, technology neutrality, collective use and license exempt assignment.

There are three models that could help understand the concept of commons (Ostrom, 1990). The first one is called the 'Tragedy of the Commons' and was first authored by Hardin (1968) to address the problem of uncontrollable population growth. He argued that, in a commons, every individual seeks to maximize his gain without recourse to a world that is limited. Hardin (1968) also argued that selling the commons as private property is essential as otherwise the destruction of the commons is inevitable.

The second model is the 'prisoner's dilemma', where each user has no incentive to cooperate with the others and always opts to leave no matter what the other users do. The third model is the 'logic of collective action', which was developed by Olson (1965) to challenge 'group theory' which suggests that individuals with commons interests in a group tend to act in support of their group objective as they would all be better off if that objectives were achieved. Olson (1965) argued that *'unless the number of individuals in quite small, or unless there is coercion or some other special device to make individuals act in their common interest, rational self-interested individuals will not act to achieve their common or group interests'*. Furthermore, Ostrom (1990) pointed out that the main common issue between the three models is the 'free rider problem' where each individual is motivated not to contribute to achieve the group objective because they cannot be excluded from the benefits that others provide.

Several scholars have addressed the concept of commons in spectrum. Gilder (1994) opposed auctioning exclusive spectrum rights as it would impede technologies that do not need exclusivity. Furthermore, he calls for handling spectrum as public property as advanced technologies will render spectrum not scarce but abundant. Baran (1995) shares a similar view and explains that spectrum is not used most of the time, and this unused spectrum could be utilized if technology could determine empty slots in terms of both time and place. The issue attracted more attention when Noam (1995, 1998) advocated the 'open access' paradigm, where users could obtain access to spectrum without having an exclusive license by buying 'access tickets' whose price varies with congestion and is automatically determined by the demand and supply conditions at the time. Noam anticipated that license exclusivity would be technologically obsolete and economically inefficient in the future, arguing that spectrum policy should instead focus on controlling the traffic among the equipment rather than controlling the spectrum (Noam, 1995, 1998). In his contribution, Benkler (1998) argued that spectrum management should focus on regulating the use of equipment rather than the spectrum in a way similar to the Internet, which adopts a decentralized commons structure. The discussion was further developed by other such as Lehr (2004) and Lehr and Crowcroft (2005), who called for a 'dedicated unlicensed spectrum' where all unlicensed devices are considered primary users will gain the full benefits of spectrum commons. Moreover, Werbach (2004) proposed a new approach for spectrum management called 'supercommons' that refocused wireless regulation away from the ownership of spectrum and towards the rights to use devices for communication.

Furthermore, it is argued that, while too many users with inclusive rights can lead to overconsumption, too many with exclusive rights to the same resource, without a clear hierarchy or dispute resolution, can lead to underconsumption or what has been called by some the 'Tragedy of the Anticommons' (Werbach, 2011). More specifically, while in the tragedy of the commons, several users have the rights to access the spectrum without excluding other (e.g. overgrazed fields), in anticommons, several users have the rights to exclude others, and no user has an effective right of use (Heller, 1998).

On the other hand, Hazlett (1998) argued that the open access model could render spectrum worthless and uninhabitable if the transaction costs of aggregating spectrum are substantial. Moreover, Hazlett (2001) criticizes spectrum commons—once an unlicensed operator creates killer applications and services, other operators will copy the business model. This may increase congestion and degrade service quality. Hazlett (2006) opposed the view that the regulator should set aside spectrum for unlicensed use to add social value, arguing that having private property right will enable the regulator to create spectrum commons. In addition, Hazlett and Leo (2010) argued that the spectrum commons approach could destroy the value of spectrum as it usually results in the distribution of a huge number of small and overlapping spectrum rights that cannot be re-aggregated. They argue also that spectrum commons is itself a rejection of open access as it might exclude and restrict certain wireless applications and services. Moreover, the lack of spectrum ownership could cause market failure as it discourages incumbents from investing in both technology and infrastructure.

It should also be noted that the term 'spectrum commons' is used by some scholars to point towards different technologies such as short-range devices, Wi-Fi, or cognitive radio systems (CRS), or to different spectrum management approaches such as unlicensed bands, overlay or underlay. However, we use the term 'spectrum commons' to signify specific spectrum management approaches that include 'spectrum open access' (Noam, 1995), 'dedicated unlicensed spectrum' (Lehr, 2004), 'private commons' (Webb & Cave, 2003), 'California dream' (Chaduc & Pogorel, 2008) and 'supercommons' (Werbach, 2004). Having said that, spectrum management approaches that accommodate unlicensed devices such as Wi-Fi or short-range devices such as RFID are usually called 'restricted commons' when the technology is standardized or 'standard commons' where technology selection is neutral (Chaduc & Pogorel, 2008).

2.5.4 Spectrum Easements

An easement is the right to use the property of another without possessing it. Easements in spectrum are largely associated with Faulhaber and Farber (2003), who suggested an approach based on allowing users other than the spectrum owner to use the spectrum as a non-interference easement. Faulhaber and Farber (2003) explain that this easement creates a non-interfering commons in all frequencies and locations. In general, there are two main types of access within spectrum easements: overlay (opportunistic) and underlay access. Overlay devices access the spectrum at the geographical, temporal or frequency gaps in the licensed user's transmission as long as they do not cause harmful interference (e.g. TV white spaces (TVWS)). Underlay access implies that a secondary user will transmit at lower power levels within the noise floor of licensed spectrum (e.g. UWB) (Cave & Webb, 2012). Underlay access entails power-related easements (Cave & Webb, 2003). For instance, short-range devices (SRD) could operate as secondary users along with the primary users as they transmit in a small bandwidth and use little power over short distances. Similarly, equipment such as ultra-wide band (UWB) can transmit over a large bandwidth using very low-power levels.

Overlay or opportunistic access is promoted by technologies such as CRS that are capable of measuring the radio environment and learning from experience in order to transmit dynamically in the temporal unused frequencies without the need of exclusive allocation (Mitola, 2000). CRS devices can dynamically adjust their operational parameters according to the operational and geographical environment (ITU-R, 2009). They can also access the infrastructure of primary users, their own infrastructure or through using an ad hoc connection (Akyildiz et al., 2008).

One of the main candidate spectrum bands for CRS operations is TVWS, which refers to the geographical interleaved vacant frequencies in TV spectrum. These frequencies were allocated for broadcasting but are not used within a particular area or frequency because of the need for a spectrum guard band and the geographical separation between TV channels to avoid interference (Freyens & Loney, 2011a). TVWS could also exist in areas where spectrum is not used for broadcasting due to limited supply of demand for broadcasting services (Freyens & Loney, 2011b). It is worth highlighting that TVWS exists regardless of whether the broadcasting services are analogue or digital. CRS usually uses technologies such as software-defined radio (SDR) to enable their behaviour to be adjusted automatically (IEEE, 2008). SDR is defined by the IEEE as a type of radio where some or all of the physical layer functions are software controlled (IEEE, 2008).

Another concept that is associated with CRS and SDR is dynamic spectrum access (DSA). This is a technique by which a radio device can dynamically select the operating spectrum it uses. DSA could be cooperative based, in which the secondary user may only use a band with permission of the primary user, or non-cooperative, where the secondary user does not require permission from the primary user (Chapin & Lehr, 2007). CRS is perceived to be an enabler of DSA that can enable a transition from a regulatory framework of spectrum allocation to spectrum access (Marshall, 2010).

One of the first steps in the implementation of CRS was when the FCC started to draw attention to the concept in 2003. It initiated a proceeding to facilitate opportunities for deploying such technologies (FCC, 2003). In 2006, the FCC allowed fixed unlicensed devices to operate in the TVWS, excluding channel 37, and prohibited personal/portable devices from operating on channels 14–20 as they are used for public safety in some cities (FCC, 2006). The FCC then decided in 2008 to allow both fixed and portable devices to operate in the white spaces on the condition that they deploy geolocation capabilities and access a database that provides a list of available TV channels that may be used at their location. (FCC, 2008).

Another development occurred in 2012 when the FCC decided to slightly increase the maximum permissible power spectral density (PSD) for each category of TV band device in order to decrease the operating cost of TVWS devices and to enable them to provide greater coverage (Yang, 2014). It is worth highlighting Marshall's (2010) argument is that while TVWS was first adopted in the USA, this may not be helpful for DSA. This is because the USA is a unique case of DSA deployment due to the existence of large areas with low populations and large amount of spectrum assigned to the military. In Europe, the Radio Spectrum Policy Group (RSPG) within the European Commission (EC) recommended that introducing CRS could be considered on the national level if it took into account border coordination issues (RSPG, 2011).

The practice of TVWS has varied around the world with limited commercial deployments and trials. In the USA, the FCC allowed fixed unlicensed devices to operate in the TVWS (FCC, 2006), and, in 2008, fixed and portable devices were allowed to operate on the condition that they deploy geolocation capabilities and access a database (FCC, 2008). In 2012, the FCC decided to slightly increase the maximum permissible power spectral density (PSD) for each category of TV bands device (Yang, 2014). In 2020, the FCC considered updates to TVWS rules in order to permit higher transmit power and antenna height above average terrain for fixed white space devices in less congested geographic areas (Alleven, 2020).

In India, eight experimental licences were issued for TVWS in the 470–582 MHz frequency band in 2016 (Times of India, 2016). Ofcom in the UK has permitted license exempt operation of TVWS in the UHF band since 2015 (Ofcom, 2016). Several Asian Pacific countries established, in April 2009, a task group on CRS and SDR to facilitate the study on these systems in their countries (APT/AWG Task Group on SDR & CRS, 2014). Finally, several TVWS initiatives have been established in Africa (e.g. Kenya) (Haji, 2014) while the ATU has established a group in 2014 to study the management of TVWS technologies (ATU, 2014).

With regard to standardization, the IEEE founded Group 802.22 in 2004 (IEEE, 2004). The group aims to develop a standard for a cognitive radio-based access interface for use by license exempt devices on a non-interfering basis in spectrum that is allocated to TV broadcast services. The IEEE 802.22–11 standard was finalized in 2011, providing broadband wireless access over a large area up to 100 km from the transmitter with a data rate up to 22 Mbps per channel without interfering with the reception of TV broadcast stations (IEEE, 2011). Moreover, the IEEE is developing IEEE 802.11af to modify the current IEEE 802.11 standard of Wi-Fi to meet the legal requirements for channel access and coexistence in TVWS (Baykas et al., 2010). The new Wi-Fi standard will enable a larger range and better indoor coverage to be achieved due to the nature of UHF spectrum.

The CRS concept has been gathering momentum and support from many organizations. For instance, the Wireless Innovation Alliance (WIA) launched a campaign in 2007 to promote the use of TVWS for wireless broadband (Digital Communities, 2007). WIA membership includes over-the-top (OTT) serviced providers, civil society and manufactures with no participation from mobile operators (Wireless Innovation Alliance, 2012). Another forum is the Whitespace Alliance (WA), which was formed at the end of 2011 (Businesswire, 2011). It focuses on the deployment and use of standards-based products and services to provide broadband wireless access in TVWS.

In addition to TVWS, manufactures, namely Qualcomm and Nokia, have proposed a new approach called ‘authorized shared access’ (ASA) to enable the dynamic overlay usage of the mobile operators’ spectrum (Standeford, 2011). ASA, it is suggested, will operate on a shared and non-interference basis in bands allocated to the mobile service and identified for IMT (ECC, 2014). It is estimated by the two companies that the assignment of 200 MHz of ASA spectrum, assuming 25 per cent occupancy by the incumbent users, could result in direct benefits of €65 billion annually (Standeford, 2011).

The ASA concept presents a further step away from CRS in TVWS. Firstly, it enables licensed secondary access rather than unlicensed. Secondly, ASA is provided by an operator and a manufacture who sells equipment capable of accessing the spectrum on an opportunistic basis. Thirdly, ASA does not provide a best effort service but instead a service with a predictable quality of service (QoS) (Qualcomm & Nokia, 2011).

ASA concept was perceived as a novel way to have two primary users with priorities instead of the traditional ITU model of primary and secondary. It would alter the traditional international allocation system. LSA is different from ASA in that sharing rules must be approved by the regulator and incorporated in operators' licenses (Forge et al., 2012). One significant advantage of LSA is that it targets spectrum that offers a clear potential for global harmonization such as the spectrum already identified for IMT (DIGITALEUROPE, 2012).

Another similar approach has been suggested in the USA by the President's Council of Advisors on Science and Technology (PCAST) (President's Council of Advisors on Science and Technology, 2012). Marcus (2012) considers that the PCAST report focused on the equivalent of LSA with a focus on the spectrum owned by the federal government being shared with private sector users. One other concept for easement is 'pluralistic licensing', where licenses are awarded under the assumption that opportunistic secondary spectrum access will be allowed (Holland et al., 2012). Faussurier (2014) notices that there is a shift in CEPT from focusing on CRS and TVWS towards spectrum sharing in general. In addition, in the Commonwealth of Independent States (CIS), there has been some work on the principles of coordination in the border areas for CRS using a national geolocation database (Kokotov, 2014). A number of Asian Pacific countries established a Task Group on CRS and SDR in April 2009 to facilitate the study on these systems (APT/AWG Task Group on SDR & CRS, 2014).

Meanwhile, several scholars have criticized the easements approach. Werbach (2004), for example, argues against deploying underlay or overlay within an exclusive property rights. First, not only could the owner prevent other users from accessing its frequencies even if they did not interfere, but it is also difficult to determine the price and terms of such access. Even if they reach an agreement with the owner, the access price could be considered as an inefficient transaction cost. Werbach (2004) further explains that the easement concept, which is based on non-interference restrictions on the secondary users, enables the primary users or the spectrum owner to raise artificial limits on the easement or to claim that interference is occurring. Furthermore, Cave and Webb (2012) argue that deploying overlay with full protection to the primary users would result in inefficiency due to the large guard bands that are required. Therefore, they suggest applying the same methodology for interference determination used for underlay to overlay along with using a database that contains maximum transmitted power for overlay users according to their position. A supplementary measure to the database is to use a spectrum monitoring system (QinetiQ, 2005). Werbach (2010) proposes a mechanism called 'spectrum networking database' (SND) to facilitate the management of dynamic access to spectrum in a way similar to the domain name system used in the management of the Internet management. Veenstra and Leonhard (2008) argue that technologies such as SRD do not fit into the present legal framework and the corresponding institutional arrangements such as certification, licensing or monitoring.

There has also been some concerns expressed regarding the different spectrum easements models. For instance, the GSMA has been sceptical about the TVWS concept for two reasons. Firstly, TVWS may disturb the existing market through inappropriate regulations such as eliminating the cost of acquiring licensed spectrum. Secondly, TVWS may have influenced the mobile allocation process to be considered at WRC-15. Thirdly, there is a lack of studies evaluating the interference effects between mobile broadband systems using TVWS and digital TV reception (GSMA, 2013). Moreover, the use of techniques like 'single frequency networks' (SFNs), where several transmitters can use the same frequency channel without interference, decreases the opportunities to use TVWS (Gomez, 2013). Forde and Doyle (2013) also clarified that there have been high expectations associated with CRS technologies that may have led to market and regulatory uncertainty. Another criticism of CRS is related to the hidden terminal problem, where CRS may identify spectrum as vacant upon measurements while the spectrum is used by another user behind an adjacent building (Webb, 2007).

Finally, studies by CEPT have shown that the amount of TVWS is limited due to digital broadcasting planning and because the TV bands are heavily used on an opportunistic basis by devices called Programme Making and Special Event (PMSE) (Anker, 2010). Having said this, this has been contradicted by other studies which show that 56% of the TV channels are unused in Europe when averaged over the whole geographic area (Cui & Weiss, 2011). In general, the amount of TVWS is highly dependent on the density of TV transmitters (Jantti et al., 2011).

2.6 Summary

This chapter has presented an overview of spectrum management at the national level from both theoretical and practical viewpoints. Firstly, it has shed light on radio spectrum as a resource, clarifying its main characteristics that make it different from other natural resources. The chapter then showed how the perception of radio spectrum has changed, from being a new inexhaustible natural resource commonly and freely used by the public to a scarce resource that needs to be managed. Although the practice of spectrum management nationally has evolved differently around the world, spectrum has been strictly regulated by most governments.

The chapter has also analysed national spectrum management along its main elements. The first element, service allocation, is related to the distribution of spectrum to the different radiocommunication services, and it could be harmonized or allowed to be flexible according to the user's demands. The second element, technology selection, could be neutral, restricted to standardized technologies or selective of specific technologies. Spectrum usage rights could be exclusive property rights, exclusive property rights with easement or collective usage rights where easements entail allowing other users to access a spectrum that is owned or licensed to a particular entity without causing harmful interference. The fourth element of spectrum management policy is frequency assignment, where there are three categories, namely administrative, market based or license exempt.

The chapter then addressed the different combinations of the four spectrum management elements which formulate national spectrum management policies. We concentrated on the four main approaches that are often considered in the literature: command and control, spectrum market, spectrum commons and spectrum easements. The command and control approach is based on allocation harmonization, standardized technology, exclusive property rights and administrative assignment. This traditional approach to spectrum management worked well for many years. It has, however, been called into question due to a combination of market liberalization, the growth in data, shortage of spectrum and the inefficiencies associated with the low utilization of spectrum. In particular, we have illustrated how the command and control was, until recently, the dominant approach. While there have been three alternatives suggested to overcome the deficiencies of such the traditional approach, none of them show wide adoption or significant success with the exception of technologies such as Wi-Fi and measures such as auctions.

The next chapter discusses in detail spectrum management internationally by analysing it in terms of the four main elements that have been used in this chapter to examine national spectrum management policies, that is, service allocation, technology selection, usage rights and frequency assignment. This analysis enables us to identify where interaction between national and international spectrum management policies and regimes occurs.

References

Agence Nationale des Fréquences (ANF). (2012). *Assignation des fréquences*. Retrieved from http://www.anf.tn.

Akalu, R., & Arias, A. D. (2012). Assessing the policy of spectrum trading in the UK. *info, 14*(1).

Akyildiz, I. F., Lee, W. Y., Vuran, M. C., & Mohanty, S. (2008). A survey on spectrum management in cognitive radio networks. *IEEE Communications Magazine, 46*(4).

Alleven, M. (2020). Microsoft's TV white space proposal moves forward. Retrieved from https://www.fiercewireless.com/.

Anker, P., & Lemstra, W. (2011). Governance of radio spectrum: License exempt devices. In W. Lemstra, V. Hayes, & J. Groenewegen (Eds.), *The innovation journey of Wi-Fi: The road to global success*. Cambridge University Press.

Anker, P. (2010). *Cognitive radio, the market and the regulator*. Paper presented at the IEEE Dyspan 2010 Conference.

APT/AWG Task Group on SDR & CRS. (2014). *APT wireless group activities on CRS & SDR*. Paper presented at the ITU-R SG 1/WP 1B Workshop: Spectrum Management issues on the use of White Spaces by Cognitive Radio Systems.

ATU. (2014). *Report of the 2nd meeting of the African spectrum working group (AfriSWoG-2)* Paper presented at the 2nd Meeting of the African Spectrum Working Group (AfriSWoG-2) Nairobi. http://www.atu-uat.org/.

Baran, P. (1995). Is the UHF frequency shortage a self made problem? In *Marconi centenniel symposium*.

Baumol, W., & Robyn, D. (2006). Toward an evolutionary regime for spectrum governance: Licensing or unrestricted entry? In *AEI brookings joint center for regulatory studies*.

Baykas, T., Wang, J., Rahman, M. A., Tran, H. N., Song, C., Filin, S., Alemseged, Y., Sun, C., Villardi, G. P., Sum, C. S., Harada, H. (2010). *Overview of TV white spaces: Current regulations, standards and coexistence between secondary users*. Paper presented at the IEEE 21st International Symposium on Personal, Indoor and Mobile Radio Communications.

Benkler, Y. (1998). Overcoming Agoraphobia: Building the commons of the digitally networked environment. *Harvard Journal of Law and Technology, 11*(2), 1–113.

Benkler, Y. (2003). The political economy of commons. *The European Journal for the Informatics Professional, IV*(3), 6–9.

Benkler, Y. (2011). *Open wireless versus licensed spectrum: Evidence from market adoption*. Draft Working Paper.

Bennett, R. (2010). Going mobile: Technology and policy issues in the mobile internet. *Information Technology and Innovation Foundation*.

Businesswire. (2011). White space alliance formed to deliver affordable, high-speed broadband internet access to 3.5 Billion Households. *Press Release*.

Cave, M. (2002). *Review of radio spectrum management, an independent review for department of trade and industry and HM treasury*. Retrieved from www.ofcom.org.uk.

Cave, M. (2006). Spectrum management and broadcasting: Current issues. *Communications and Strategies, 62*, 19–34.

Cave, M. (2008). Market-based methods of spectrum management in the UK and the European Union. *Telecommunications Journal of Australia, 58*(2/3), 1–24.

Cave, M., & Webb, W. (2012). The unfinished history of usage rights for spectrum. *Telecommunications Policy, 36*(4), 293–300.

Cave, M., & Foster, A. (2010). Solving spectrum gridlock: Reforms to liberalize radio spectrum management in Canada in the face of growing scarcity. *C.D. Howe Institute, 303*.

Cave, M., & Morris, A. (2005). *Getting the best out of public sector spectrum*. Paper presented at the The 33th Annual Telecommunications Policy Research Conference.

Cave, M., & Valletti, T. (2000). Are spectrum auctions ruining our grandchildren's future? *Info, 2*(4), 347–350.

Cave, M., & Webb, W. (2003). Designing property rights for the operation of spectrum markets. *Papers in Spectrum Trading* (1).

Cave, M., Foster, A., & Jones, R. W. (2006). *Radio spectrum management: Overview and trends*. Paper presented at the ITU Workshop on Market Mechanisms for Spectrum Management. http://www.itu.int.

Cave, M., Doyle, C., & Webb, W. (2007). *Essentials of modern spectrum management*. Cambridge University Press.

CEPT. (2010). *CEPT report 39: Report from CEPT to the European Commission in response to the mandate to develop least restrictive technical conditions for 2 GHz bands*. Retrieved from http://www.erodocdb.dk.

Chaduc, J., & Pogorel, G. (2008). *The radio spectrum. Managing a strategic resource*. ISTE Ltd.

Chapin, J. M., & Lehr, W. H. (2007). The path to market success for dynamic spectrum access technology. *IEEE Communications Magazine, 45*(5), 96–103.

Cisco. (2012). *Cisco visual networking index: Global mobile data traffic forecast update, 2011–2016*. Retrieved from http://www.cisco.com.
Coase, R. H. (1959). The Federal Communications Commission. *Journal of Law & Economics, 2*(1), 1–40.
Coase, R. H. (1960). The problem of social cost. *Journal of Law & Economics, 3*(Oct), 1–44.
Coase, R. H. (1998). Comment on Thomas W. Hazlett: Assigning property rights to radio spectrum users: Why did FCC license auctions take 67 years? *Journal of Law & Economics, 41*(2), 577–580.
Cramton, P. (2002). Spectrum auctions. In M. Cave, S. Majumdar, & I. Vogelsang (Eds.), *Handbook of Telecommunications economics*. Elsevier Science B.V.
Cui, L., & Weiss, M. B. (2011). *Can unlicensed bands be used by unlicensed usage?* Paper presented at the TPRC41: The 41st Research Conference on Communication, Information and Internet Policy.
Dictionaries, A. H. (2001). Spectrum. In *The American heritage dictionary of the english language* (4th edition ed.). Houghton Mifflin Company.
DIGITALEUROPE. (2012). *Position paper on the European Commission Communication "promoting the shared use of spectrum in the internal market" COM(2012) 478*. Retrieved from http://www.digitaleurope.org/.
Digital Communities. (2007). Wireless innovation alliance launches campaign to bring broadband access to all americans. *News Reports*.
ECC. (2014). *ECC report 205: Licensed shared access (LSA)*. Retrieved from http://www.erodocdb.dk/.
Economist, T. (2004). On the same wavelength. *The Economist* (Special Report Spectrum Policy).
Electronic Communications Committee (ECC). (2011). *Description of practices relative to trading of spectrum usage rights*.
El-Moghazi, M., Whalley, J., & Irvine, J. (2012). *WRC-12: Implication for the spectrum eco-system*. Paper presented at the TPRC42: The 42nd Research Conference on Communication, Information and Internet Policy.
ERC. (1998). *ERC report 53: Report on the introduction of economic criteria in spectrum management and the principles of fees and charging in the CEPT*.
European Commission. (1998). *COM (98) 596 final: Green paper on radio spectrum policy*. Retrieved from https://eur-lex.europa.eu.
Eurostrategies and LS-Telecom. (2007). *Study on radio interference regulatory models in the European community*.
Falch, M., & Tadayoni, R. (2004). Economic versus technical approaches to frequency management. *Telecommunications Policy, 28*, 197–211.
Faulhaber, G. R., & Farber, D. J. (2003). *Spectrum management: Property rights, markets, and the commons*. AEI-Brookings Joint Center for Regulatory Studies, Working Paper 02–12.
Faussurier, E. (2014). *Introduction of new spectrum sharing concepts: LSA and WSD*. Paper presented at the ITU-R SG 1/WP 1B Workshop: Spectrum Management issues on the use of White Spaces by Cognitive Radio Systems.
FCC. (2002). *Report of the spectrum policy task force*. Retrieved from www.fcc.gov.
FCC. (2003). *Facilitating opportunities for flexible, efficient, and reliable spectrum use employing cognitive radio technologies, authorization and use of software defined radios, ET Docket No. 03–108, ET Docket No. 00–47, FCC 03–322*.
FCC. (2006). *Unlicensed operation in the TV broadcast bands, first report and order and further notice of proposed rule making, ET Docket No. 04–186, ET Docket No. 02–380, FCC 06–156*.
FCC. (2008). *Unlicensed operation in the TV broadcast bands, second report and order and memorandum opinion and order, ET Docket No. 04–186, ET Docket No. 02–380, FCC 08–2360*.
Forde, T., & Doyle, L. (2013). A TV whitespace ecosystem for licensed cognitive radio. *Telecommunications Policy, 37*(2–3), 130–139.
Forge, S., Horvitz, R., & Blackman, C. (2012). *Perspectives on the value of shared spectrum access: Final report for the European Commission*. Retrieved from http://ec.europa.eu.

Foster, A., Cave, M., & Jones, R. W. (2011). Going mobile: Managing the spectrum. In C. Blackman & L. Srivastava (Eds.), *Telecommunications regulation handbook*. The World Bank.
Foster, A. (2008). *Spectrum sharing*. Paper presented at the GSR 2008.
Freyens, B. P. (2009). A policy spectrum for spectrum economics. *Information Economics and Policy, 21*, 128–144.
Freyens, B. P. (2010). Shared or exclusive radio waves? A dilemma gone astray. *Telematics and Informatics, 27*(3), 293–304.
Freyens, B. P., & Loney, M. (2011a). *Digital switchover and regulatory design for competing white space usage rights*. Paper presented at the IEEE Dyspan 2011 Conference.
Freyens, B. P., & Loney, M. (2011b). Projecting regulatory requirements for TV white space devices. In S. J. Saeed & R. A. Shellhammer (Eds.), *TV white space spectrum technologies, regulations, standards, and applications*. CRC Press.
Frullone, M. (2007). *A deeper insight in technology and service neutrality*. Paper presented at the ITU Workshop on Market Mechanisms for Spectrum Management.
Gilder, G. (1994). Auctioning the airways. *Forbes*.
Gomez, C. (2013). *TV white spaces: Managing spaces or better managing inefficiencies?* Paper presented at the GSR 2013.
Goodman, E. P. (2004). Spectrum rights in the telecosm to come. *San Diego Law Review, 41*(269), 269–404.
GSMA. (2013). *GSMA position on TV white spaces*. Retrieved from http://www.gsma.com.
Haji, M. A. (2014). *Licensing of TV white space networks in Kenya*. Paper presented at the ITU-R SG 1/WP 1B Workshop: Spectrum Management issues on the use of White Spaces by Cognitive Radio Systems.
Hardin, G. (1968). The tragedy of the commons. *Science, 162*(3859), 1243–1248.
Hazlett, T. W. (1998). Spectrum flash dance: Eli Noam's proposal for "open access" to radio waves. *Journal of Law & Economics, 41*(2), 805–820.
Hazlett, T. W. (2001). The wireless craze, the unlimited bandwidth myth, the spectrum auction faux pas, and the punchline to ronald coase's "Big joke": An essay on airwave allocation policy. *Harvard Journal of Law & Technology, 13*(2), 332–567.
Hazlett, T. W. (2006). The spectrum-allocation debate: An analysis. *IEEE Internet Computing, 10*(5), 68–74.
Hazlett, T. W., & Leo, E. T. (2010). The case for liberal spectrum licenses: A technical and economic perspective. George Mason Law & Economics Research Paper No. 10–19.
Hazlett, T. W., & Muñoz, R. E. (2006). Spectrum allocation in Latin America: An economic analysis. *George Mason Law & Economics Research Paper, 6*(44), 261–278.
Heller, M. (1998). The tragedy of the anticommons: Property in the transition from marx to markets. *Harvard Law Review, 111*(3), 621–688.
Herzel, L. (1952). [Facing facts about the broadcast business]: Rejoinder. *The University of Chicago Law Review, 20*(1), 106–107.
Herzel, L. (1998). My 1951 color television article. *Journal of Law & Economics, 41*(2), 523–527.
Herzel, L. (1951). 'Public interest' and the market in color television regulation. *University of Chicago Law Review*. 802–816.
Holland, O., Nardis, L. D., Nolan, K., Medeisis, A., Anker, P., Minervini, L. F., Velez, F., Matinmikko, M., Sydor, J. (2012). *Pluralistic licensing*. Paper presented at the IEEE Dyspan 2012 Conference.
Horvitz, R. (2013). Geo-database management of white space versus open spectrum. In E.Pietrosemoli & M.Zennaro (Eds.), *Tv white space: A pragmatic approach*. ICTP.
IEEE. (2004). IEEE Starts standard to tap open regions in the TV spectrum for wireless broadband services. *News Release*. Retrieved from http://www.ieee.org.
IEEE. (2008). IEEE standard definitions and concepts for dynamic spectrum access: Terminology relating to emerging wireless networks, system functionality, and spectrum management.
IEEE. (2011). IEEE 802.22–2011 standard for wireless regional area networks in TV whitespaces completed. *Press Release*. Retrieved from http://www.ieee802.org.

Indepen and Aegis Systems. (2004). *Costs and benefits of relaxing international frequency harmonisation and radio standards*. Retrieved from stakeholders.ofcom.org.uk.
Indepen, A. (2001). *Implications of international regulation and technical considerations on market mechanisms in spectrum management: Report to the independent spectrum review*. Retrieved from http://www.ofcom.org.uk.
ITU-R. (2001a). ITU-R recommendation SM.1132–2: General principles and methods for sharing between radiocommunication services or between radio stations. In *SM series. Spectrum Management*. ITU.
ITU-R. (2001b). ITU-R recommendation SM.1265–1: National alternative allocation methods. In *SM series. Spectrum management*. ITU.
ITU-R. (2004). *ITU-R report SM.2012–2. Economic aspects of spectrum management*. ITU.
ITU-R. (2008). Article 1: Terms and definitions. In *Radio regulations*. ITU.
ITU-R. (2009). ITU-R report SM.2152. Definitions of software defined radio (SDR) and cognitive radio system (CRS). In *SM series. Spectrum management*. ITU.
ITU-R. (2020). Article 1: Terms and definitions. In *Radio regulations*. ITU.
Jantti, R., Kerttula, J., Koufos, K., & Ruttik, K. (2011). *Aggregate interference with FCC and ECC white space usage rules: Case study in Finland*. Paper presented at the IEEE Dyspan 2011 Conference.
Kelly, R. B., & LaFrance, A. J. (2012). Spectrum trading in the EU and the US—Shifting ends and means. In R. Bratby (Ed.), *The international comparative legal guide to: Telecommunication laws and regulations 2012*: Global Legal Group.
Kokotov, O. (2014). *Cognitive radio systems. Principles of coordination in the border areas*. Paper presented at the ITU-R SG 1/WP 1B Workshop: Spectrum Management issues on the use of White Spaces by Cognitive Radio Systems.
Kruse, E. (2002). From free privilege to regulation: Wireless firms and the competition for spectrum rights before world war I. *The Business History Review, 76*(4), 659–703.
Kwerel, E., & Williams, J. (2002). A proposal for a rapid transition to market allocation of radio spectrum. *FCC OPP Working Paper, 38*.
Lehr, W. (2005). *The role of unlicensed in spectrum reform*. Massachusetts Institute of Technology.
Lehr, W., & Crowcroft, J. (2005). *Managing shared access to a spectrum commons*. Paper presented at the IEEE Symposium on New Frontiers in Dynamic Spectrum Access Networks.
Lehr, W. (2004). Dedicated lower-frequency unlicensed spectrum: The economic case for dedicated unlicensed spectrum below 3 GHz. *New America Foundation, Spectrum Policy Program, Spectrum Series Working Paper 9*.
London Economics. (2008). *Economic Impacts of increased flexibility and liberalisation in European spectrum management: Report for A group of European communications sector companies*. Retrieved from http://londoneconomics.co.uk/.
Marcus, M. J. (2012). Spectrum sharing issues on both sides of the Atlantic. *Wireless Communications, IEEE, 19*(6), 6–7.
Marcus, B. K. (2004). The spectrum should be private property: The economics, history, and future of wireless technology. *Essays In political economy, Ludwig von Mises Institute*.
Marshall, P. F. (2010). *A potential alliance for World-Wide dynamic spectrum access*. Paper presented at the IEEE Symposium on New Frontiers in Dynamic Spectrum Access Networks.
McAfee, P., McMillan, J., & Wilkie, S. (2010). The greatest auction in history. In J. J. Siegfried (Ed.), *Better living through economics*. Harvard University Press.
Mitola, J. (2000). *Cognitive radio: An integrated agent architecture for software defined radio*. (Ph.D.). Royal Institute of Technology.
Netting, R. (1976). What alpine peasants have in common: Observations on communal tenure in a Swiss village. *Human Ecology, 4*(2), 135–146.
Noam, E. (1995). Taking the next step beyond spectrum auctions—Open spectrum access. *IEEE Communications Magazine, 33*(12), 66–73.

Noam, E. (1998). Spectrum auctions: Yesterday's heresy, today's orthodoxy, tomorrow's anachronism. Taking the next step to open spectrum access. *Journal of Law & Economics, 41*(2), 765–790.

Nordicity. (2010). *Study on the market value of fixed and broadcasting satellite spectrum in Canada.* Retrieved from https://www1.ic.gc.ca/.

OECD. (2006). *The spectrum dividend: Spectrum management issues.* Retrieved from www.oecd.org.

Ofcom. (2012). *Securing long term benefits from scarce spectrum resources. A strategy for UHF bands IV and V*. Retrieved from http://stakeholders.ofcom.org.uk.

Ofcom. (2016). *TV White Spaces.* Retrieved from http://stakeholders.ofcom.org.uk.

Olson, M. (1965). *The logic of collective action. Public goods and the theory groups.* Harvard University Press.

Oranje, C. V., Cave, J., Mandele, M. V. D., Schindler, R., Hong, S. Y., Iliev, I., & Vogelsang, I. (2008). Responding to convergence: Different approaches for telecommunication regulators. Retrieved from www.opta.nl.

Ostrom, E. (1990). *Governing the commons: The evolution of institutions for collective action.* Cambridge University Press.

Pogorel, G. (2007). Opinion: The nine regimes of spectrum management. *PolicyTracker.com.*

President's Council of Advisors on Science and Technology (PCAST). (2012). Realizing the full potential of government-held spectrum to spur economic growth.

QinetiQ. (2005). *Bandsharing concepts.* Retrieved from Farnborough. www.spectrumaudit.org.uk.

Qualcomm, & Nokia. (2011). *Authorised shared access: An evolutionary spectrum authorisation scheme for sustainable economic growth and consumer benefit*. Paper presented at the The 72th Working Group Frequency Management Meeting Miesbach. www.cept.org.

Ratto-Nielsen, J. (2006). *The international telecommunications regime: Domestic preferences and regime change*: lulu.com.

Rothbard, M. N. (1982). Law, property rights, and air pollution. *Cato Journal, 2*(1).

RSPG. (2011). *RSPG opinion on cognitive technologies.* Retrieved from http://rspg.groups.eu.int.

Smythe, D. W. (1952). Facing facts about the broadcast business. *The University of Chicago Law Review, 20*(1), 96–106.

Standeford, D. (2011). Qualcomm and Nokia propose authorised shared access to spectrum. *PolicyTracker.com.*

Struzak, R. (2003). *Introduction to international radio regulations.* Retrieved from Italy.

The New America Foundation and The Shared Spectrum Company. (2003). *Dupont circle spectrum utilization during peak hours.* Retrieved from http://www.newamerica.net.

Times of India. (2016). Government allocates 127 MHz spectrum for TV white space technology test. *Times of India.*

Timofeev, V. (2006). From radiotelegraphy to worldwide wireless., how ITU processes and regulations have helped shape the modern world of radiocommunications. *ITU News, 3*, 5–9.

Veenstra, Y., & Leonhard, H. (2008). *Avoiding harmful interference—The current paradigm.* Paper presented at the The Annual Conference on Competition and Regulation on Network Industries (CRNI).

Vogler, J. (2000). *The global commons: Environmental and technological governance.* Wiley-Blackwell.

Webb, W., & Cave, M. (2003). Spectrum licensing and spectrum commons—Where to draw the line. *Papers in Spectrum Trading,* (2).

Webb, W. (2007). *Wireless communications: The future.* Wiley.

Wellenius, B., & Neto, I. (2005). The radio spectrum: Opportunities and challenges for the developing world. *World Bank Policy Research Working Paper, 3742.*

Werbach, K. (2004). Supercommons: Toward a unified theory of wireless communication. *Texas Law Review, 82*, 863–973.

Werbach, K. (2010). Castle in the air: A domain name system for spectrum. *Northwestern University Law Review, 104*(2), 613–640.

Werbach, K. (2011). The Wasteland: Anticommons, white spaces, and the fallacy of spectrum. *Arizona Law Review, 53*(1), 213–254.

Whittaker, M. (2002). *True technology neutral spectrum licences.* Retrieved from www.futurepace.com.au.

WIK. (2006). *Towards more flexible spectrum regulation.* Retrieved from www.wik.com.

Wireless Innovation Alliance. (2012). Members. (1/5/2012). Retrieved from https://www.wirelessinnovationalliance.com.

Yang, A. (2014). *Overview of FCC's new rules for TV white space devices and database updates.* Paper presented at the ITU-R SG 1/WP 1B Workshop: Spectrum Management issues on the use of White Spaces by Cognitive Radio Systems.

Chapter 3
International Radio Spectrum Management

All stations, whatever their purpose, must be established and operated in such a manner as not to cause harmful interference to the radio services or communications of other Members or of recognized operating agencies, or of other duly authorized operating agencies which carry on a radio service, and which operate in accordance with the provisions of these Regulations. The application of the provisions of these Regulations by the International Telecommunication Union does not imply the expression of any opinion whatsoever on the part of the Union concerning the sovereignty or the legal status of any country, territory or geographical area.

The RR preamble (2020)

3.1 Introduction

This quote is the heart of the international radio regulations (RRs). On the one hand, each country has the right to utilize its radio spectrum in any way it chooses but, on the other hand, countries should not cause interference to the stations of another country that are operated according to the RRs. In other words, there are tensions between the applicability of the RRs within a country and between neighbouring countries. This is one of the several paradoxes within the RRs that this book seeks to highlight.

Before exploring the tensions that exist within the RRs, this chapter will investigate how spectrum is managed internationally in order to highlight the main elements of the interaction between international and national spectrum management policies and regimes. The early development of spectrum management internationally and how it was created in parallel with the formulation process of national spectrum management policies are analysed. In particular, two issues necessitated managing the spectrum internationally: the phenomenon of interference and the refusal of Marconi to relay messages received from competing operators.

The chapter also questions the main principles of spectrum management internationally. It starts with brief background on the ITU as an international organization

M. A. El-Moghazi and J. Whalley, *The International Radio Regulations*,
https://doi.org/10.1007/978-3-030-88571-7_3

and then focuses on the ITU-R to trace the origins of international spectrum management. Following this, the chapter then explores the RR and its main elements before highlighting the different ITU-R recommendations and resolutions and other documentations that are accepted and adopted by many countries despite not being part of the RR. Finally, the chapter concludes with a summary.

3.2 The International Telecommunication Union

The origin of the ITU can be traced back to the issue of incompatibility of telegraph systems among countries that led to the first International Telegraph Convention, which was held in Paris in 1865. At that time, the International Telegraph Union (the first incarnation of the ITU) was established to supervise subsequent amendments to the agreement that would standardize telegraph equipment. In 1885, the International Telegraph Conference was held in Berlin to draw up international legislation governing telephony (ITU, 2020).

In the 1890s, many inventors worked on transferring signals over the air, and wireless telegraphy was possible. Similar to wireline telegraph, there were issues of incompatibility between the wireless systems being adopted in different countries. The issue was apparent when Prince Henry of Prussia, returning across the Atlantic from a visit to the USA, attempted to send a courtesy message from his ship to US President Theodore Roosevelt in 1902. However, the message was declined by the US shore station because the ship's radio equipment was of a different type and nationality from that onshore.

This led Germany to call a Preliminary Radio Conference in Berlin in 1903 to establish international regulations for radiotelegraph communications. This was followed a few years later by the first International Radiotelegraph Conference in 1906 in Berlin, which decided that the Bureau of the ITU would act as the conference's central administrator and to establish the Radiotelegraph Section of the Bureau. Most importantly, the conference issued the International Radiotelegraph Convention with an annex containing the first form of the RR as they exist today and established 'SOS' as the international maritime distress call.

The conference also established the principle of compulsory communication between land and vessels at sea (Timofeev, 2006). The conference decision was a reflection of the issue of multisystem interoperability that emerged when Marconi equipment operators were refusing to relay messages received from competing operators (Anker & Lemstra, 2011). Following the Titanic tragedy, the International Radiotelegraph Conference, held in London in 1912, agreed on a common wavelength for the radio distress signals of ships.

In 1932, it was decided to change the name of the 'International Telegraph Union' to its current name of 'International Telecommunication Union', and the International Telegraph Convention of 1865 was combined with the International Radiotelegraph Convention of 1906 to form the International Telecommunication Convention. In 1947, the ITU was recognized as the UN specialized agency for telecommunications.

The highest management body of the ITU is the Plenipotentiary Conferences (PP) which are held every four years to revise the ITU constitution and convention. The Plenipotentiary Conference (PP), in Geneva in 1992, structured the ITU into three sectors: Telecommunication Standardization (ITU-T), Radiocommunication (ITU-R) and Telecommunication Development (ITU-D). Currently, the ITU membership includes 193 Member States in addition to Sector Members from the private sector. Between the consecutive PPs, the ITU council is responsible of policy and budget decisions and is composed of one-quarter of the ITU Member States (ITU-R, 2015a).

The ITU has an important role in the different wireless industries. For instance, the ITU's first technical standards for television were released in 1949, and, during the space age, the ITU held an Extraordinary Administrative Conference for space communications in 1963. The ITU also has an important role with regard to global flight tracking when the ITU PP of 2014 approved Resolution 185 that instructed WRC-15 to include in its agenda the consideration of the issue. That was in response to the loss of Malaysian Airlines Flight MH 370 (ITU, 2014). As a result, WRC-15 allocated the 1087.7–1092.3 MHz frequency band to the aeronautical mobile-satellite service (earth-to-space) for reception by space stations of Automatic-Dependent Surveillance-Broadcast (ADS-B) emissions from aircraft transmitters (ITU, 2015).

In addition, the ITU has reached many decisions that greatly impacted on the daily lives of people around the world. Examples include but are not limited to allocating the 5150–5350 MHz and 5470–5725 MHz bands for mobile services to enable Wi-Fi systems to utilize it (WRC-03), allocating additional spectrum for IMT-2000 systems (WRC-2000), replanning the broadcasting-satellite service (WRC-97) and allocating additional spectrum to the mobile-satellite service (MSS) (Timofeev, 2006).

The ITU constitution is the main instrument that sets out the basic principles of the organization, and the ITU Convention complements the constitution, providing detailed information on its functionality (ITU-R, 2018). The preamble of the ITU constitution highlights that ITU Member States have agreed to it in order to facilitating peaceful relations, international cooperation among peoples and economic and social development by means of efficient telecommunication services while fully recognizing the sovereign right of each country to regulate its telecommunication (ITU, 2019). Article 1 of the constitution highlights the purposes of the ITU, which includes maintaining and extending international cooperation among all its Member States for the improvement and the rational use of telecommunications of all kinds, aligning the actions of Member States and promoting and offering technical assistance to developing countries in the field of telecommunications.

The article also clarifies the actions taken by the ITU to achieve these purposes, which include:

- Effect allocation of bands of the radio-frequency spectrum, the allotment of radio frequencies and the registration of radio-frequency assignments and, for space services, of any associated orbital position in the geostationary-satellite orbit or of any associated characteristics of satellites in other orbits, in order to avoid harmful interference between radio stations of different countries.

- Coordinate efforts to eliminate harmful interference between radio stations of different countries and to improve the use made of the radio-frequency spectrum for radiocommunication services and of the geostationary-satellite and other satellite orbits.
- Facilitate the worldwide standardization of telecommunications, with a satisfactory quality of service.
- Foster international cooperation and solidarity in the delivery of technical assistance to the developing countries and the creation, development and improvement of telecommunication equipment and networks in developing countries by every means at its disposal, including through its participation in the relevant programmes of the United Nations and the use of its own resources, as appropriate (ITU, 2019).

Article 5 explains that the provisions of the constitution and convention are complemented by the two binding administrative regulations: International Telecommunication Regulations (ITRs) and the radio regulations (RR) (ITU, 2019). The ITRs have a similar binding status to the RR and they are the result of the merger of the International Telegraph Regulations and Telephone Regulations in 1988. These regulations were revised in 2012 at the World Conference on International Telecommunications (WCIT-12) (ITU, 2020).

Article 7 explores the structure of the ITU, which is composed of the Plenipotentiary Conference, the Council, World Conferences On International Telecommunications, the Radiocommunication Sector, the Telecommunication Standardization Sector, the Telecommunication Development Sector, and the General Secretariat (ITU, 2019).

The constitution accommodates important provisions for radio spectrum. Most importantly, Article 44 'Use of the Radio-Frequency Spectrum and of the Geostationary-Satellite and Other Satellite Orbits' states that '*Member States shall endeavour to limit the number of frequencies and the spectrum used to the minimum essential to provide in a satisfactory manner the necessary services. To that end, they shall endeavour to apply the latest technical advances as soon as possible* (ITU, 2019). The article also states that '*In using frequency bands for radio services, Member States shall bear in mind that radio frequencies and any associated orbits, including the geostationary-satellite orbit, are limited natural resources and that they must be used rationally, efficiently and economically, in conformity with the provisions of the radio regulations, so that countries or groups of countries may have equitable access to those orbits and frequencies, taking into account the special needs of the developing countries and the geographical situation of particular countries*' (ITU, 2019).

Another important provision is Article 45 'Harmful Interference', which states that '*All stations, whatever their purpose, must be established and operated in such a manner as not to cause harmful interference to the radio services or communications of other Member States or of recognized operating agencies, or of other duly authorized operating agencies which carry on a radio service, and which operate in accordance with the provisions of the radio regulations*'. Article 46 is related to distress calls and messages, instructing that '*Radio stations shall be obliged to accept, with absolute priority, distress calls and messages regardless of their origin, to reply in the same manner to such messages, and immediately to take such action in regard thereto as may be required*'.

3.3 The Radio Sector of ITU

It was the Radiotelegraph Conference of 1927 that established the International Radio Consultative Committee (CCIR), which was many years later in 1992 became the ITU-R (Ard-Paru, 2013). The first time that spectrum was allocated in response to a need rather than legalizing existing uses was in 1938 at the Cairo Radio Conference. In addition, the Cairo Conference enhanced spectrum utilization efficiency by establishing technical standards and restricting the use of frequency wasteful for radio transmitters (Codding, 1991). The PP of 1932 divided the world for frequency allocation purposes into two regions, Europe and elsewhere, and established two technical tables, one of the frequency tolerances and the other of acceptable bandwidths (ITU-R, 2015a). In the following years, the interest of many stakeholders in spectrum, such as the military and broadcasters, started to grow, which, in turn, motivated governments to heavily regulate spectrum usage (Ryan, 2012). This was in line with the national policies at the turn of the twentieth century, which sought to enforce government control over radio spectrum for political purposes such as national security and for economic benefits derived from controlling the industry (Cowhey, 1990).

Figure 3.1 shows the organizational structure of the ITU-R (ITU, 2021). The ITU-R Bureau (BR) is responsible of providing administrative and technical support, recording and registering frequency assignments and allotments and also orbital characteristics of space stations, and maintaining the Master International Frequency Register (MIFR). It can also provide advice to Member States and investigate and assist in resolving cases of harmful interference (ITU-R, 2015a). The ITU-R BR accommodates four departments: Space Services (SSD), Terrestrial Services (TSD), Study Groups (SGD) and Informatics, Administration and Publications (IAP).

One important body within the ITU-R is the Radio Regulations Board (RRB). The RRB is composed of 12 elected members, and it is responsible for approving the Rules of Procedure and addressing matters which cannot be resolved through the application of the RR and Rules of Procedure. According to Article 13 of the RR, the RRB will develop a new Rule of Procedure only when there is a clear need with proper justification for such a Rule, which is then submitted to the next WRC to make the necessary modifications to the RR (ITU-R, 2020k) The RRB also handles reports of unresolved interference investigations and can provide advice to Radiocommunication Conferences. Previously, the RRB operated under the name of International Frequency Registration Board (IFRB), which was established by the PP of 1947 and placed in charge of the Master of Frequency list. The IFRB developed into the RRB in 1992 (Ard-Paru, 2013).

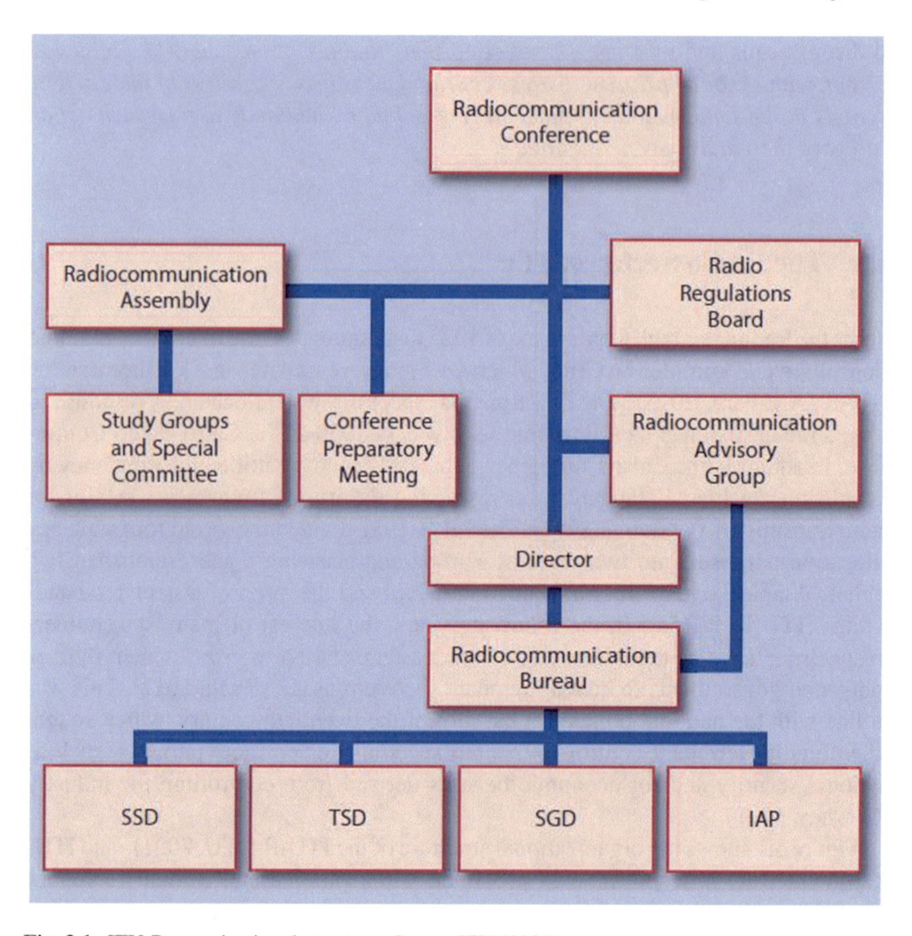

Fig. 3.1 ITU-R organizational structure. *Source* ITU (2021)

The second important body is the World Radiocommunication Conferences which revises the RR, frequency plans and tackle issues that are of global or ITU regional interest. WRC are conducted every three or four years and their agenda is set by the Council on the basis of the draft agenda agreed by the previous WRC. Conference Preparatory Meetings (CPM) prepare a consolidated report on the technical, operational and regulatory and procedural bases for the forthcoming WRC. Another important type of conference is the Regional Radiocommunication Conferences (RRCs) that deal with radiocommunication issues relevant to a particular region (region) and the needs of its Member States.

The Radiocommunication Assembly (RA) is responsible for the structure, programme and approval of radiocommunication studies conducted by the ITU-R study groups and is usually aligned with WRCs. The ITU study groups were originally established in 1948 as CCIR study groups as follows:

- SG I Radio transmitters
- SG II Radio receivers
- SG III Complete radio systems employed by the different services
- SG IV Ground wave propagation
- SG V Tropospheric propagation
- SG VI Ionospheric propagation
- SG VII Radio time signals and standard frequencies
- SG VIII International monitoring
- SG IX General technical questions
- SG X Broadcasting including questions related to single sideband
- SG XI Television including questions related to single sideband
- SG XII Tropical broadcasting
- SG XIII Operation questions depending principally on technical considerations.

The structure of the study groups was further modified before reaching its current form:

- SG 1 Spectrum utilization—monitoring
- SG 2 Space research and radioastronomy services
- SG 3 Fixed service at frequencies below about 30 MHz
- SG 4 Fixed service using satellites
- SG 5 Propagation in non-ionized media
- SG 6 Ionospheric propagation
- SG 7 Standard frequency and time-signal services
- SG 8 Mobile services
- SG 9 Fixed service using radio-relay systems
- SG 10 Broadcasting service (sound)
- SG 11 Broadcasting service (television) (ITU, 1979).

Currently, there are six study groups (SGs): SG1 (Spectrum management), SG3 (Radiowave propagation), SG4 (Satellite services). SG5 (Terrestrial services), SG6 (Broadcasting service) and SG7 (Science services). These SGs develop ITU-R Recommendations, Reports and Handbooks and accommodate Working Parties that focus on specific issues.

Another important body within the ITU-R is the Radiocommunication Advisory Group (RAG) which reviews the priorities and strategies adopted in the ITU-R. Conference Preparatory Meeting (CPM) prepares a consolidated report on the technical, operational and regulatory and procedural bases for the forthcoming WRC (ITU-R, 2015a).

3.4 The Radio Regulations

The RR, or the 'red book' as it is sometimes called, are much more than just the international table of service allocations. The RR are more than 2000 pages of articles, resolutions, recommendations, appendices, and ITU-R recommendations incorporated by reference. The first volume starts with the Preamble of the RR, which accommodates the main principles of the RR that are also part of the ITU constitution as follows:

- '*Members shall endeavour to limit the number of frequencies and the spectrum used to the minimum essential to provide in a satisfactory manner the necessary services. To that end, they shall endeavour to apply the latest technical advances as soon as possible*' (No. 195 of the Constitution of the International Telecommunication Union (Geneva, 1992)).
- '*In using frequency bands for radio services, Members shall bear in mind that radio frequencies and any associated orbits, including the geostationary-satellite orbit are limited natural resources and that they must be used rationally, efficiently and economically, in conformity with the provisions of these Regulations, so that countries or groups of countries may have equitable access to those orbits and frequencies, taking into account the special needs of the developing countries and the geographical situation of particular countries*' (No. 196 of the Constitution).
- '*All stations, whatever their purpose, must be established and operated in such a manner as not to cause harmful interference to the radio services or communications of other Members or of recognized operating agencies, or of other duly authorized operating agencies which carry on a radio service, and which operate in accordance with the provisions of these Regulations*' (No. 197 of the Constitution).

Generally speaking, these principles focus on specific issues such as using the spectrum efficiently without hoarding it using modern technologies. These principles also highlight an important concept which is equality in accessing the spectrum considering the needs of developing countries. Not causing harmful interference is also at the heart of these regulations. The preamble also states the objectives of the RR:

- '*To facilitate equitable access to and rational use of the natural resources of the radio-frequency spectrum and the geostationary-satellite orbit.*
- *To ensure the availability and protection from harmful interference of the frequencies provided for distress and safety purposes.*
- *To assist in the prevention and resolution of cases of harmful interference between the radio services of different administrations.*
- *To facilitate the efficient and effective operation of all radiocommunication services.*
- *To provide for and, where necessary, regulate new applications of radiocommunication technology*' (ITU-R, 2020n).

The last paragraph of the preamble states that *'The application of the provisions of these Regulations by the International Telecommunication Union does not imply the expression of any opinion whatsoever on the part of the Union concerning the sovereignty or the legal status of any country, territory or geographical area'*. In other words, the application of the RR should not contradict the sovereignty of those countries who agree to be bound by the RR.

Volume 1 of the RR contains 58 articles with the most important one of them being Article 5 'frequency allocations', which is almost 150 pages in length. Article 1 accommodates terms and definitions related to issues such as radio services, radio stations and systems and characteristics of emissions and radio equipment. For instance, radio is defined as *'general term applied to the use of radio waves'*, and radiocommunication is defined as '*telecommunication by means of radio waves*' (ITU-R, 2020h).

One section of Article 1, 'frequency sharing', contains definitions of interference and its forms. Firstly, interference is defined as '*The effect of unwanted energy due to one or a combination of emissions, radiations, or inductions upon reception in a radiocommunication system, manifested by any performance degradation, misinterpretation, or loss of information which could be extracted in the absence of such unwanted energy*'. Permissible interference is defined as '*observed or predicted interference which complies with quantitative interference and sharing criteria contained in the RR or in ITU-R Recommendations or in special agreements*' and acceptable interference is defined as *'Interference at a higher level than that defined as permissible interference and which has been agreed upon between two or more administrations without prejudice to other administrations'*.

Finally, harmful interference is defined as '*interference which endangers the functioning of a radio navigation service or of other safety services or seriously degrades, obstructs, or repeatedly interrupts a radiocommunication service operating in accordance with the RR*'. A first observation that can be made is that the definition of harmful interference is also mentioned and used in the ITU constitution. Secondly, according to the RR, not all forms of interference are considered harmful; interference is permitted if its value is below the values specified within ITU-R recommendations. In other cases where interference values are above those mentioned in the recommendations but allowed by administration, it is considered acceptable interference. Meanwhile, some elements of the 'harmful interference' definition such as 'degrades' are not quantified and it is not clear whether the degradation is due to inefficiencies of the receiver or transmitter characteristics.

Article 4 is related to the assignment and use of frequencies, and Article 4.1 is similar to the principle in the RR preamble calling for limiting the number of frequencies and the spectrum used to the minimum essential to provide in a satisfactory manner the necessary services while utilizing the latest technical advances. A well-known provision of Article 4 is 4.4, which provides the flexibility for administrations to use the spectrum in degradation of the RR on the condition of not causing harmful interference to, or claiming protection from, harmful interference caused by a station operating in accordance with the RR (ITU-R, 2020i).

Article 5 is the heart of the RR and presents the frequency allocations in the 8.3 kHz–3000 GHz bands. In many instances in Article 5, there are footnotes that point to other resolutions and appendixes in the other volumes of the RR. For instance, Article 5.132 points to Appendix 17 with respect to the international frequencies for the transmission of maritime safety information (MSI), and within Article 5.132A applications of radiolocation services are limited to oceanographic radars operating in accordance with Resolution 612 (Rev.WRC-12) (ITU-R, 2020m).

There are other articles related to the notification, coordination and registration of frequencies in the ITU-R Master International Frequency Register such as Articles 8, 9 and 11. Most importantly, Article 8.1 states that '*The international rights and obligations of administrations in respect of their own and other administrations' frequency assignments1 shall be derived from the recording of those assignments in the Master International Frequency Register (the Master Register) or from their conformity, where appropriate, with a plan. Such rights shall be conditioned by the provisions of these Regulations and those of any relevant frequency allotment or assignment plan*' (ITU-R, 2020j).

Article 15, section I 'interference from radio stations', provides some details regarding how to decrease or prevent interference. There are provisions that request transmitting stations to limit their radiation power as much as possible to ensure a satisfactory service and request also to take advantage of directional antennas whenever the nature of the service permits while minimizing out of band emissions. Article 15, section VI 'procedure in a case of harmful interference' provides detailed information on the required steps to report and resolve harmful interference. According to Article 15.27, as much detail as possible should be provided in the form indicated in Appendix 10. Moreover, Article 15.34 states that '*Having determined the source and characteristics of the harmful interference, the administration having jurisdiction over the transmitting station whose service is being interfered with shall inform the administration having jurisdiction over the interfering station, giving all useful information in order that this administration may take such steps as may be necessary to eliminate the interference*' (ITU-R, 2020l).

The second volume of the RR contains appendices that complement the articles of the first volume. For instance, Article 3.9 addresses the technical characteristics of stations with a focus on the bandwidths of emission and it points to Appendix 1 'Classification of emissions and necessary bandwidths' as follows: '*Appendix 1 is provided as a guide for the determination of the necessary bandwidth*' (ITU-R, 2020a). Furthermore, Article 15.27, within the context of reporting harmful interference, points to forms within Appendix 10 (ITU-R, 2020b). Appendix 18 includes the table of transmitting frequencies in the VHF maritime mobile band as incited in Article 52 (ITU-R, 2020c). Similarly, Appendix 27 contains the frequency allotment plan for the aeronautical mobile (R) service and related information as referenced in Article 43 (ITU-R, 2020d).

Another part of the Appendix details the different planned services (e.g. aeronautical, maritime). For example, Appendix 30 lists broadcasting-satellite services in the 11.7–12.2 GHz bands in Region 3 and 11.7–12.5 GHz bands in Region 1 (ITU-R, 2020e), whereas Appendix 30B details fixed-satellite services in a number of frequency bands including 4500–4800 MHz (ITU-R, 2020g). These appendices contain several articles that provide details on the execution, notification and modification of these plans.

The third volume of the RR accommodates WRC resolutions and recommendations, which are usually the results of specific agenda items or proposals at WRCs. While both resolutions and recommendations have a binding status as they are part of the RR, resolutions contain stronger language with Member States under a greater obligation to follow them. It is also difficult to define categories for these resolutions and recommendations as they are related to different issues. Some of them determine rules within the RR, for example, Resolution 26 specifies rules for the usage of footnotes to the Table of Frequency Allocations in Article 5 of the RR (ITU-R, 2020r).

Other resolutions have a political nature such as Resolution 12 which states '*that assistance to Palestine, pursuant to the relevant ITU resolutions and decisions shall be continued, in particular through capacity building, with the view to enabling Palestine to obtain and manage the required radio spectrum in order to operate its telecommunication networks and wireless services*' (ITU-R, 2020p). Moreover, resolutions such as Resolution 7 address spectrum management nationally in terms of the establishment and operation of radio-frequency management units (ITU-R, 2020o). Another focus of WRC resolutions is related to the needs of developing countries. For instance, Resolution 20 addresses the technical cooperation with developing countries in the field of aeronautical telecommunications (ITU-R, 2020q).

In some cases, WRC resolutions are necessary to regulate the sharing between different services. For instance, Resolution 752 determines the sharing criteria between active and passive services in the 36–37 GHz band (ITU-R, 2020w). Similarly, Resolution 760 provides provisions relating to the use of the 694–790 MHz frequency band in Region 1 by mobile services, except aeronautical, and by other services (ITU-R, 2020x). Resolution 750 is quite critical for the compatibility between the earth exploration-satellite service (passive) and relevant active services where unwanted emissions of stations brought into use in the frequency bands and services listed in the resolution shall not exceed the limits mentioned in the resolution (ITU-R, 2020v).

One of the most important resolutions for the satellite industry is Resolution 86, which invites '*future World Radiocommunication Conferences to consider any proposals which deal with deficiencies and improvements in the advance publication, coordination, notification and recording procedures of the RR for frequency assignments pertaining to space services which have either been identified by the Board and included in the Rules of Procedure or which have been identified by administrations or by the Radiocommunication Bureau, as appropriate*' (ITU-R, 2020t). The resolution is a standing item on WRC agendas to address improvements in the satellite regulatory procedures including filings to facilitate rational efficient and economical access to spectrum for satellites (Ofcom, 2016).

There are also several resolutions that address the future need of specific systems such as IMT and HAPS prior to WRC as part of the conference agenda items. For instance, Resolution 160 provides detailed information on the studies to be conducted by the ITU-R regarding additional spectrum needs for gateway and fixed terminal links for HAPS as part of Agenda Item 1.14 of WRC-19 (ITU-R, 2020u). Similarly, Resolution 238 of WRC-15 provided a framework for sharing and compatibility studies as part of AI 1.13 of WRC-19 (ITU-R, 2016a). Another type of resolution is the one that determines the agenda items for the two next WRCs. For instance, Resolution 809 of WRC-15 provided the agenda for the WRC-19 (ITU-R, 2016b), while Resolution 810 outlined the preliminary agenda for WRC-23 (ITU-R, 2016c).

In order to understand how the ITU-R draft WRC resolutions, it is perhaps useful to dissect one of these resolutions in order to understand its structure. For instance, Resolution 234, which reviews the use of spectrum in the 470–960 MHz frequency band in Region 1, determines the framework for studies on the UHF under AI 1.5 of WRC-23. It starts with a part labelled 'considering' that provides background to the resolution such as that the favourable propagation characteristics in the frequency bands below 1 GHz being beneficial to providing cost-effective solutions for coverage, and that the 470–862 MHz frequency band is a harmonized band used to provide terrestrial television broadcasting services on a worldwide scale.

The second part is 'recognizing' which provides more factual information such as the GE06 Agreement applies in all Region 1 countries, except Mongolia, and in Iran (Islamic Republic of), for the 470–862 MHz frequency band. Another part is 'noting' which has a relatively similar function to 'considering'. The most important part of the resolution is the 'resolve' part that determines the details of the framework for the required studies. Resolution 234 invites ITU-R to review the spectrum use and study the spectrum needs of existing services within the 470–960 MHz frequency band in Region 1 taking into account relevant ITU-R studies, recommendations and reports. The last part, 'invite', is dependent on the context of the resolution and varies according to the concerned parties with respect to the resolution issue (e.g. ITU-D, administrations).

The fourth volume of the RR—'ITU-R Recommendations incorporated by reference'—accommodates the recommendation issued by the Radio Assembly that have a relatively binding nature and are considered as part of the RR treaty unlike other ITU-R recommendations. One recommendation incorporated by reference is ITU-R M.1652-1. This recommendation—'Dynamic frequency selection (DFS) in wireless access systems including radio local area networks for the purpose of protecting the radiodetermination service in the 5 GHz band'—was approved by the RA of 2003 and incorporated by reference into the RR by WRC 2003 (ITU-R, 2011). The recommendation provides requirements for dynamic frequency selection (DFS) as a mitigation technique to be implemented in wireless access systems (WAS) including radio local area networks (RLANs) to facilitate sharing with the radiodetermination service in the 5 GHz band.

The approval of such a recommendation by RA-03 facilitated the discussion of an additional mobile allocation in the 5 GHz to introduce wireless access systems (WAS) including radio local area networks (RLANs). More specifically, changing the legal status of such a recommendation to be binding provided certainty for radar users and facilitated reaching an agreement on the issue during WRC-03. The conference subsequently resolved in Resolution 229 to allocate the band between 5150 and

5350 MHz and between 5470 and 5725 MHz on a coprimary basis to '*Wireless Access Systems including RLANs. It also resolves that in the bands 5250–5350 MHz and 5470–5725 MHz, the mitigation measures found in Annex 1 to Recommendation ITU-R M.1652-1 shall be implemented by systems in the mobile service to ensure compatible operation with radiodetermination system*' (ITU-R, 2003b).

WRC resolution 27 clarified the principles of the use of incorporation by reference into the RR. The resolution noted that the principles of incorporation by reference were firstly adopted by WRC-95, and that there are provisions in the RR containing references which fail to adequately distinguish whether the status of the referenced text is mandatory or non-mandatory. The resolution resolved that the term 'incorporation by reference' should only apply to those references intended to be mandatory. The annex to Resolution 27 clarified that where the relevant texts are brief, the referenced material should be placed in the body of the RR rather than using incorporation by reference and that the text incorporated by reference shall have the same treaty status as the RR themselves (ITU-R, 2020s).

3.5 ITU-R Resolutions and Recommendations

In the previous section, we examined the different WRC resolutions and recommendations that are part of the RR and have treaty status. We also highlighted the ITU-R recommendations incorporated by reference, which are issued by the Radio Assembly but exceptionally have a binding nature. The other ITU-R resolutions and recommendations do not have a binding status and are not part of the RR. They are, however, accepted and followed by most countries. More specifically, it is quite important for administrations and the telecommunications industry to have their requirements and operational conditions included in these documents.

While it is usually argued that the ITU-R is only concerned with international spectrum issues, this is not always the case. In particular, there are a number of ITU-R recommendations and reports that address national issue. For instance, the latest version of the Handbook on National Spectrum Management, which is from 2015, includes information relevant for countries on issues such as licensing, monitoring and assignment (ITU-R, 2015a). It also includes examples of multilateral agreements for frequencies assignment outside of the ITU-R framework such as the Vienna agreement among European countries that addresses the coordination of frequencies between 29.7 MHz and 43.5 GHz for fixed services and land mobile services.

In addition, ITU-R resolutions provide instructions on the organization, methods or programmes of the RA and SGs. One of the most important resolutions is ITU-R resolution 1 which is entitled 'Working methods for the Radiocommunication Assembly, the Radiocommunication Study Groups, the Radiocommunication Advisory Group and other groups of the Radiocommunication Sector'. Its latest version is Resolution 1-8 (2019). While the '1' indicates the resolution number, the '8', indicates the number of versions following the revisions of the resolution and the year, '2019', is the year when the RA approved the resolution or its revised version.

Resolution 1 describes the procedures for adoption, approval and suppression of the different ITU-R documents. A draft revised or new resolution is needed to be adopted within each SG by consensus among all those Member States who attend and are approved by the RA (ITU-R, 2019a).

The second form of ITU-R documentation is ITU-R questions, which are defined as '*a statement of a technical, operational or procedural study, generally seeking a Recommendation, Report or Handbook*' (ITU-R, 2019a). There are different categories of questions assigned to SG. More specifically, there are two categories for questions:

- 'C', conference-oriented questions associated with the work related to the preparations and decisions of World and Regional Radiocommunication Conferences
- 'S', questions which are intended to respond to matters referred to the Radiocommunication Assembly by the Plenipotentiary or any other conference, the Council or the Radio Regulations Board. The questions may also reflect advances in radiocommunication technology or spectrum management, as well as operational or usage changes.

There are two categories of 'C' questions, namely 'C1', which are those very urgent and priority studies required for the next World Radiocommunication Conference, and 'C2', urgent studies, which are required for other radiocommunication conferences. There are also three categories of 'S' questions: 'S1', urgent studies to be completed within two years, 'S2', important studies, necessary for the development of radiocommunications, and 'S3', required studies expected to facilitate the development of radiocommunications (ITU-R, 2019b).

An example of a question is ITU-R 240/1 'Assessment of spectrum efficiency and economic value'. In this question, the RA decided that the following questions should be studied: what method should quantify spectrum efficiency, which factors define the economic value of spectrum, and is there a general model to assess the economic value of spectrum? The RA also decided that the results of the above studies should be included in Recommendation(s), Report(s) and Handbook(s), with all of them completed by 2023. The question is assigned the category of 'S2', indicating the priority and urgency allocated to the questions to be studied (ITU-R, 2017b).

An answer to a question or parts of a question could be in the format of a ITU-R Recommendation which '*provides recommended specifications, requirements, data or guidance for recommended ways of undertaking a specified task; or recommended procedures for a specified application, and which is considered to be sufficient to serve as a basis for international cooperation in a given context in the field of radiocommunications*' (ITU-R, 2019a). The process of adoption and approval of ITU-R recommendations is quite sophisticated and involves different procedures according to the circumstances. However, the approval process generally has two main stages. The first one is the adoption by the SG, while the second is the approval by the Member States, either by consultation between RAs or at an RA.

During the SG meeting, if there is no objection by any administration attending the meeting to the new or revised recommendation, its approval simultaneously occurs by what is called the 'PSAA procedure' (i.e. the 'procedure for simultaneous adoption and approval'). In such cases, that recommendation will be circulated to all Member States and Sector Members, and, if no objection is received from a Member State, the draft new or revised Recommendation is then considered to be adopted by the SG. This clearly indicates that only a single administration can stop the approval of a recommendation. For example, ITU-R recommendation ITU-R M 1036 states the different frequency arrangements for the implementation of the terrestrial component of IMT in the bands identified in the RR (ITU-R, 2015b). A recommendation structure is similar to ITU-R resolution where it has 'recognizing', 'noting' and 'considering' parts and then a 'recommend' section instead of 'resolve'. There are different series of ITU-R recommendations according to their focus. For instance, the 'BT' series is for broadcasting services, 'F' series for fixed services and 'M' for mobile services.

A SG could prepare an ITU-R report instead of a recommendation in response to a question, which involves less sophisticated procedures for approval, and, in case one administration does not approve the report, an attributed statement in the report and/or in the summary record of the SG meeting can be included. Similar to the ITU-R recommendation, reports have different series (e.g. 'M', 'F'). A report is different from a recommendation where the former accommodates informative data such as guidelines, examples and case studies that do not recommend or support a specific technology or measure.

Another key ITU-R document is a handbook that is defined as '*A text which provides a statement of the current knowledge, the present position of studies, or of good operating or technical practice, in certain aspects of radiocommunications, which should be addressed to a radio engineer, system planner or operating official who plans, designs or uses radio services or systems, paying particular attention to the requirements of developing countries*' (ITU-R, 2019a). Over the years, ITU-R has issued several handbooks (ITU-R,).

Finally, ITU-R Opinion is defined as '*A text containing a proposal or a request destined for another organization (such as other Sectors of ITU, international organizations,* etc.*) and not necessarily relating to a technical subject*'' An example of ITU-R opinions is Opinion 103 where ITU- R SG 6 is of the opinion that '*citizens of the world would benefit if manufacturers of mobile telephones, tablets, and similar devices as well as associated service providers would include and activate a broadcast radio tuner functionality in their products along with the appropriate applications to facilitate radio broadcasting reception*' (ITU-R, 2017a).

3.6 Summary

In this chapter, we have addressed the different aspects of international spectrum management with a focus on the ITU-R and the international RR. The origin of the ITU was related to the importance of international legislation governing telegraphy and telephony systems in the nineteenth century. Similar to wireline telegraph, there were two issues that necessitated coordinating wireless systems and managing the radio spectrum internationally: the phenomenon of interference and the refusal of

Marconi to relay messages received from competing operators at the beginning of the twentieth century. In 1947, the ITU was recognized as the UN specialized agency for telecommunications, with the highest management body being the Plenipotentiary Conferences (PP). This is held every four years to revise the ITU constitution and convention.

The ITU-R is the administrative cooperation body responsible for setting the international spectrum management rules through the RR and ITU-R resolutions, recommendations and reports, while the ITU-R BR is mainly responsible of recording and registering frequency assignments and allotments and maintaining the MIFR. The RRB addresses matters which cannot be resolved through application of the RR and Rules of Procedure. WRC revise the RR, frequency plans and tackle issues that are of global or ITU regional interest, while the RA is responsible for the structure, programme and approval of radiocommunication studies conducted by the ITU-R study groups. The RAG reviews the priorities and strategies adopted in the ITU-R, with the CPM preparing a consolidated report on the technical, operational and regulatory and procedural bases for the forthcoming WRC.

The RR are based on limiting the number of frequencies and the spectrum used to the minimum essential to provide, in a satisfactory manner, the necessary services and not causing harmful interference to other countries or recognized operating agencies. Volume 1 of the RR contains 58 articles with the most important one being Article 5 that accommodates the international frequency allocations. The second volume of the RR contains appendices that complement the articles while the third volume covers WRC resolutions and recommendations, which are usually the results of specific agenda items or proposals at WRCs. The fourth volume includes the recommendations issued by the Radio Assembly that have a binding nature and are considered to be part of the RR treaty unlike other ITU-R recommendations.

References

Anker, P., & Lemstra, W. (2011). Governance of radio spectrum: License exempt devices. In W. Lemstra, V. Hayes, & J. Groenewegen (Eds.), *The innovation journey of Wi-Fi: The road to global success*. Cambridge University Press.

Ard-Paru, N. (2013). *Implementing spectrum commons: Implications for Thailand*. Chalmers University of Technology. PhD Thesis.

Codding, G. A. (1991). Evolution of the ITU. *Telecommunications Policy, 15*(4), 271–285.

Cowhey, P. F. (1990). The international telecommunications regime: The political roots of regimes for high technology. *International Organization, 44*(2), 169–199.

ITU. (1979). CCIR 50 Anniversary. Retrieved September 20, 2020 from https://www.itu.int

ITU. (2014). *Resolution 185: Global flight tracking for civil aviation* PP-14, Busan. ITU.

ITU. (2015). *Radio spectrum allocated for global flight tracking*. Retrieved September 9, 2020 from https://www.itu.int

ITU. (2019). *Constitution of the International Telecommunication Union* (Collection of the Basic Texts of the ITU Adopted by the Plenipotentiary Conference (Ed 2019), ITU.

ITU. (2020). *Overview of ITU's History*. Retrieved September 9, 2020 from https://www.itu.int

ITU. (2021). *ITU-R Sector Organization*. Retrieved April 14, 2021 from https://www.itu.int

ITU-R. (2003a). *Deployment of IMT-2000 systems handbook*. ITU.
ITU-R. (2003b). Resolution 229: Use of the Bands 5 150–5250 MHz, 5250–5350 MHz and 5470–5725 MHz by the mobile service for the implementation of wireless access systems including radio local area networks. In *Provisional Final Acts - World Radiocommunication Conference (WRC-2003)*.
ITU-R. (2011). ITU-R Recommendation M.1652-1. Dynamic frequency selection in wireless access systems including radio local area networks for the purpose of protecting the radiodetermination service in the 5 GHz Band. In *M Series. Mobile, Radiodetermination, Amateur and Related Satellite Services*. ITU.
ITU-R. (2015a). *Handbook on national spectrum management*. ITU.
ITU-R. (2015b). *M.1036-5 (10/2015): Frequency arrangements for implementation of the terrestrial component of International Mobile Telecommunications (IMT) in the bands identified for IMT in the Radio Regulations*. ITU.
ITU-R. (2016a). Resolution 238: Studies on frequency-related matters for International Mobile Telecommunications identification including possible additional allocations to the mobile services on a primary basis in portion(s) of the frequency range between 24.25 and 86 GHz for the future development of International Mobile Telecommunications for 2020 and beyond. In *Radio Regulations*. ITU.
ITU-R. (2016b). Resolution 809: Agenda for the 2019 World Radiocommunication Conference. In *Radio Regulations*. ITU.
ITU-R. (2016c). WRC-15 Resolution 810: Agenda for the 2023 World Radiocommunication Conference. In *Radio Regulations*. ITU.
ITU-R. (2017a). Opinion 103: The activation of radio broadcasting receivers in smart/mobile telephones and tablets. In *ITU-R Opinions*. ITU.
ITU-R. (2017b). Question 240/1: Assessment of spectrum efficiency and economic value. In *ITU-R Questions*. ITU.
ITU-R. (2018). *Report SM.2093: Guidance on the regulatory framework for national spectrum management*, SM Report Series. ITU.
ITU-R. (2019a). Resolution 1-8: Working methods for the Radiocommunication Assembly, the Radiocommunication Study Groups, the Radiocommunication Advisory Group and other groups of the Radiocommunication Sector. In *ITU-R Resolutions*. ITU.
ITU-R. (2019b). Resolution 5-8 (2019): Work programme and Questions of Radiocommunication Study Groups. In *ITU-R Resolutions*. ITU.
ITU-R. (2020a). Appendix 1: (REV.WRC-19) Classification of emissions and necessary bandwidths. In *Radio Regulations*. ITU.
ITU-R. (2020b). Appendix 10: Report of harmful interference. In *Radio regulations* (Vol. 1). ITU.
ITU-R. (2020c). Appendix 18: (REV.WRC-19) Table of transmitting frequencies in the VHF maritime mobile band. In *Radio Regulations*. ITU.
ITU-R. (2020d). Appendix 27: (REV.WRC-19) Frequency allotment Plan for the aeronautical mobile (R) service and related information. In *Radio Regulations*. ITU.
ITU-R. (2020e). Appendix 30: (REV.WRC-19) Provisions for all services and associated Plans and List for the broadcasting-satellite service in the frequency bands 11.7–12.2 GHz (in Region 3), 11.7–12.5 GHz (in Region 1) and 12.2–12.7 GHz (in Region 2) In *Radio Regulations*. ITU.
ITU-R. (2020f). Appendix 30A: (REV.WRC-19) Provisions and associated Plans and List for feeder links for the broadcasting-satellite service (11.7–12.5GHz in Region 1, 12.2–12.7 GHz in Region 2 and 11.7–12.2 GHz in Region 3) in the frequency bands 14.5–14.8 GHz and 17.3–18.1 GHz in Regions 1 and 3, and 17.3–17.8 GHz in Region 2. In *Radio Regulations*. ITU.
ITU-R. (2020g). Appendix 30B: (REV.WRC-19) Provisions and associated Plan for the fixed-satellite service in the frequency bands 4 500–4 800 MHz, 6 725–7 025 MHz, 10.70–10.95 GHz, 11.20–11.45 GHz and 12.75–13.25 GHz. In *Radio Regulations*. ITU.
ITU-R. (2020h). Article 1: Terms and Definitions. In *Radio Regulations*. ITU.
ITU-R. (2020i). *Article 4: Assignment and Use of Frequencies*. In *Radio Regulations*. ITU.

ITU-R. (2020j). Article 8: Status of frequency assignments recorded in the Master International Frequency Register. In *Radio Regulations*. ITU.

ITU-R. (2020k). Article 13: Instructions to the Bureau. In *Radio Regulations*. ITU.

ITU-R. (2020l). Article 15: Interferences. In *Radio Regulations*. ITU.

ITU-R. (2020m). Article 5: Frequency Allocations. In *Radio Regulations*. ITU.

ITU-R. (2020n). Preamble. In *Radio Regulations*. ITU.

ITU-R. (2020o). Resolution 7: (Rev.WRC-19) Development of national radio-frequency management. In *Radio Regulations*. ITU.

ITU-R. (2020p). Resolution 12: (Rev.WRC-19) Assistance and support to Palestine. In *Radio Regulations*. ITU.

ITU-R. (2020q). Resolution 20: (Rev.WRC-03) Technical cooperation with developing countries in the field of aeronautical telecommunications In *Radio Regulations*. ITU.

ITU-R. (2020r). Resolution 26: (Rev.WRC-19) Footnotes to the Table of Frequency Allocations in Article 5 of the Radio Regulations. In *Radio Regulations*. ITU.

ITU-R. (2020s). Resolution 27: (Rev.WRC-19) Use of incorporation by reference in the Radio Regulations. In *Radio Regulations*. ITU.

ITU-R. (2020t). Resolution 86: (Rev.WRC-07) Implementation of Resolution 86 (Rev. Marrakesh, 2002) of the Plenipotentiary Conference. In *Radio Regulations*. ITU.

ITU, Geneva ITU-R. (2020u). Resolution 160: (WRC-15) Facilitating access to broadband applications delivered by high-altitude platform stations. In *Radio Regulations*. ITU.

ITU-R. (2020v). Resolution 750: (Rev.WRC-19) Compatibility between the Earth exploration-satellite service (passive) and relevant active services. In *Radio Regulations*. ITU.

ITU-R. (2020w). Resolution 752: (WRC-07) Use of the frequency band 36–37 GHz. In *Radio Regulations*. ITU.

ITU-R. (2020x). Resolution 760: (Rev.WRC-19) Provisions relating to the use of the frequency band 694–790 MHz in Region 1 by the mobile, except aeronautical mobile, service and by other services. In *Radio Regulations*. ITU.

Ofcom. (2016). *UK Report on the outcome of the World Radiocommunication Conference 2015 (WRC-15)*. http://ofcom.org.uk

Ryan, P. S. (2012). The ITU and the Internet's Titanic Moment. *Stanford Technology Law Review, 8*, 1–36.

Timofeev, V. (2006). From Radiotelegraphy to Worldwide Wireless, How ITU Processes and Regulations Have Helped Shape the Modern World of Radiocommunications. *ITU News*(3). Retrieved September 30, 2012, from https://www.itu.int

Chapter 4
Radiocommunication Service Allocation

> *If the right to use a frequency is to be sold, the nature of that right would have to be precisely defined. In defining property rights, it would be necessary to take into account the existence of international agreements on the use of radio frequencies.*
>
> Coase (1959)

4.1 Introduction

Defining spectrum property rights with usage flexibility is a key issue in spectrum markets. One of the dimensions of flexibility is to have liberalization or neutrality in radiocommunication service allocation, where within the same bands several radio-communication services could be deployed (e.g. fixed, mobile and broadcasting). The main argument in favour of flexibility is that service harmonization could lead to restrictions in the use of under- or unused spectrum for alternative uses and on the ability to refarm spectrum for new services.

Internationally, there has only been a few attempts to bring greater flexibility into the international radiocommunication service allocation framework including reorganizing the radio regulations (RR), introducing a new composite service or changing service definitions. Quite simply, significant resistance from most of the international stakeholders has been encountered by all of these attempts. International coordination could be a constraint on the ability of a single country to introduce more flexibility into its spectrum use. In particular, applying flexible spectrum use that is not in conformity with the RR would require extensive coordination with neighbouring countries.

In this chapter, we address the issue of radiocommunication service allocation flexibility. The chapter starts with an exploration of radiocommunication service allocation and the different types of services (Sect. 4.2). It then addresses the concept of service allocation flexibility and its applicability. Section 4.4 then addresses the different elements of flexibility within the RR, and Sect. 4.5 focuses on flexibility within the planned services and the a priori planning concept. This is followed in Sect. 4.6 by recounting the different proposals for flexibility in service allocation and

M. A. El-Moghazi and J. Whalley, *The International Radio Regulations*,
https://doi.org/10.1007/978-3-030-88571-7_4

in the RR more generally during the last WRCs. Finally, the chapter addresses one of the main proposals for flexibility in spectrum allocation in CEPT countries, namely Wireless Access Policy for Electronic Communication Services (WAPECS).

4.2 Service Allocation

In general, the principles underpinning service allocation have not been changed since the first Radio Telegraph Conference in Berlin in 1906 (ITU-R, 2012g). In particular, dividing spectrum according to the type of service and global harmonization of spectrum allocations is the historical methods used in the ITU to mitigate harmful interference. Service allocation is also essential for interference management because these service definitions provide information on the application types and generic technical characteristics of the stations under each service type. It is worth noting, however, that the level of protection required by one type of service may not be suitable for another type of service (ITU-R, 1995b).

Before exploring the concept of service allocation in more detail, it is useful to understand the perspective of the RR on radiocommunication services. Article 1.6 defines radiocommunications as telecommunications by means of radio waves, while Article 1.19 defines radiocommunication service as a service involving the transmission, emission and/or reception of radio waves for specific telecommunication purposes (ITU-R, 2020a). In addition, allocation is defined in Article 1.16 as *'Entry in the Table of Frequency Allocations of a given frequency band for the purpose of its use by one or more terrestrial or space radiocommunication services or the radio astronomy service under specified conditions'* (ITU-R, 2020a).

As this chapter will address terrestrial services with a focus on fixed, mobile and broadcasting services, it is important to highlight the RR definitions of these services and their associated stations. Firstly, a fixed service is defined as a radiocommunication service between specified fixed points, while a fixed station is simply defined as a station in the fixed service. A mobile service is a radiocommunication service between mobile and land stations or between mobile stations, while a land station is not intended to be used while in motion. A mobile station is intended to be used while in motion or during halts at unspecified points. Finally, a broadcasting service is a radiocommunication service in which the transmissions are intended for direct reception by the general public. This service may include sound and television transmissions. A broadcasting station is a station in the broadcasting service.

Allocating spectrum to the different services is the main responsibility of the ITU-R. Article 5 of the RR divides the frequency band from 8.3 kHz to 3000 GHz into smaller bands that are allocated to more than 40 different radiocommunication services (ITU-R, 2020b). Currently, there are three types of radiocommunication services: broadly defined services (e.g. fixed, mobile), narrowly defined services (e.g. land mobile, maritime mobile) and normally defined services (e.g. broadcasting, amateur) (ITU-R, 1995b).

The origin of radiocommunication services can be traced back to 1906 when the first radiocommunication service was established under the 'maritime service' term in 1906 (Ard-Paru, 2012). By the mid-1920s, the issue of harmful interference had become more apparent as Europe and North America contained many broadcasting stations using unregulated frequencies (Savage, 1989). Accordingly, the 1927 International Radiotelegraph Conference witnessed a shift from focusing on the regulation of radio traffic towards allocating the spectrum to separate services (Woolley, 1995). Significantly, the concept of the 'common use of common frequencies' was gathering momentum (Levin, 1971).

The ITU-R divides the world into three regions in terms of spectrum allocation as shown in Fig. 4.1 (ITU-R, 2020b). The regional system was initiated in 1938 when the world was divided into Europe and 'other regions' (Savage, 1989). This was further developed in 1947 into three regions system accommodating Europe, Africa and the USSR in Region 1, the Western Hemisphere in Region 2 and the rest of the world in Region 3 (Mazar, 2009). The main reasons of having such system was the differences between the spectrum policies of the USA and European countries and the ITU voting system that favoured European countries (Ryan, 2012).

The RR differentiate between these three regions and other regions by having a capital 'R' for them for the purpose of frequency allocation. Region 1 includes Africa and Europe and the Middle East. In the RR, it is defined as the area limited on the east by Line A and on the west by line B, excluding any of the territory of the Islamic Republic of Iran which lies between these limits. Region 2 includes the area limited on the east by line B and on the west by line C, whereas Region 3 includes the area limited on the east by line C and on the west by line A except any of the territory

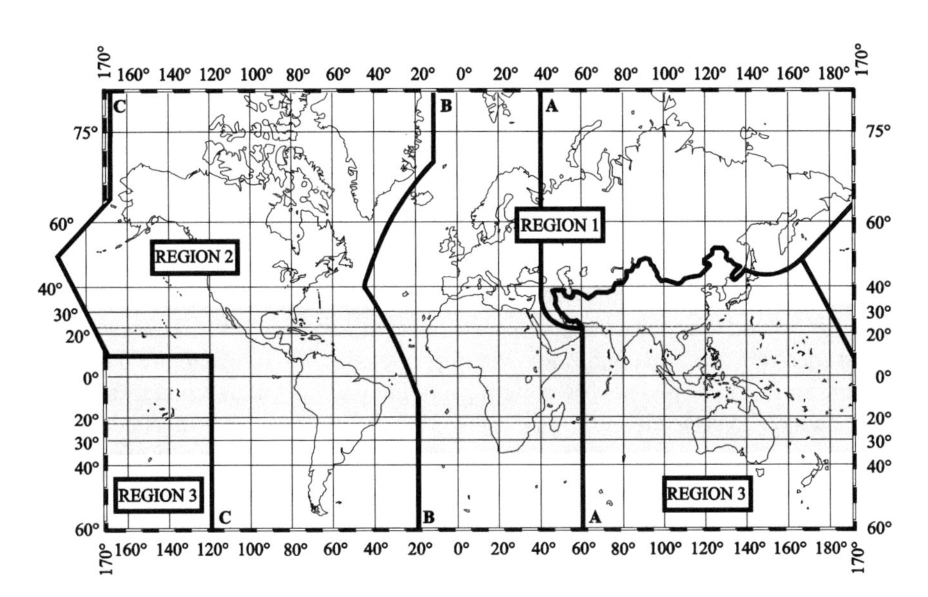

Fig. 4.1 Three regions of service allocation. *Source* ITU (2020b)

of Armenia, Azerbaijan, the Russian Federation, Georgia, Kazakhstan, Mongolia, Uzbekistan, Kyrgyzstan, Tajikistan, Turkmenistan, Turkey and Ukraine and the area to the north of Russian Federation. It also includes that part of the territory of the Islamic Republic of Iran lying outside of those limits. Lines A, B and C are defined in the RR Article 5 (ITU-R, 2020b).

In general, each spectrum band could be allocated to one or more radio services with equal or different rights (primary and secondary). This is based on the results of compatibility and sharing studies that are usually technology dependent (Louis, 2011). Article 5 clarifies the different categories of services and allocations. For instance, according to Article 5.33, where a band is indicated in a footnote of the table as allocated to a service 'on a primary basis', in a geographical area less than an ITU Region or in specific countries, this is a primary service only in that area or these countries. Similar provisions apply for secondary services. Furthermore, the order of the services allocated in the same frequency band is defined as follows: the name of the service which are in 'capitals' (e.g. FIXED) is called 'primary' services, while the names of service in 'normal characters' (e.g. mobile) are called 'secondary' services. Moreover, any additional remarks are printed in normal characters (e.g. MOBILE except aeronautical mobile).

Operations of secondary services are organized according to Article 5.28 as follows: '*Stations of a secondary service shall not cause harmful interference to stations of primary services to which frequencies are already assigned or to which frequencies may be assigned at a later date; cannot claim protection from harmful interference from stations of a primary service to which frequencies are already assigned or may be assigned at a later date; can claim protection, however, from harmful interference from stations of the same or other secondary service(s) to which frequencies may be assigned at a later date*' (ITU-R, 2020b). Article 5 clarifies cases of additional and alternative allocations. More specifically, additional allocation is an allocation via a footnote which is added in this area or in this country to the service or services which are indicated in the table of allocation. In contrast, an alternative allocation via a footnote is an allocation which replaces, in this area or in this country, the allocation indicated in the table of allocation (ITU-R, 2020b).

In cases where spectrum use nationally is in conformity with the primary allocation in the ITU-R RR, such use is protected against interference from primary and secondary services allocated in the same band or from secondary services (Indepen, 2001). It cannot, however, claim protection from primary services in the same band. If that spectrum use is in conformity with the secondary services in the ITU table of spectrum allocation, this use is protected against interference from secondary services but cannot claim protection from primary services in the same band (Indepen, 2001). Likewise, a country cannot claim protection from primary or secondary services in the same band but could operate a service on a non-protection non-interference basis according to the ITU-R Article 4.4. In addition, the country should not clash with the spectrum allocation of neighbouring countries (Foster et al., 2011).

If a country wants to protect this new use against interference, they may add a footnote to the ITU-R RR stating their particular use. However, other neighbouring countries may block this use if it restricts their own usage. Footnotes indicate additional or alternative allocations (ITU-R, 1995a). More specifically, a band could be indicated in a footnote of the ITU allocation table as being allocated to a particular service on a primary or secondary basis in a particular country or area smaller than an ITU region. Footnotes can also determine services priority and technical conditions associated with allocation (Ard-Paru, 2012).

In addition to footnotes, countries can register their important frequency assignments with the ITU. There are two types of registration (Ryan, 2005). The first is called 'a priori planning', which enables the guaranteeing of access to the spectrum where each country submits its requirements at a global or regional planning conference. An example of this approach occurred at the 2006 ITU Regional Radio Conference (RRC-06). This conference planned to allocate the 470–862 MHz band for digital broadcasting in Europe, Africa, the Middle East and Iran in what is called the 'Geneve-06 agreement' (GE-06) (O'Leary et al., 2006). The second type of assignment registration within the non-planned bands is on a 'first come, first-served' basis in the ITU-R Master International Frequency Register (MIFR). This, however, may stand in contrast with the preamble of the ITU Constitution that recognizes the sovereign right of each member state to regulate its telecommunications so that each country has the freedom to assign its frequencies to any service or station and acquire international protection required registering those frequencies in the ITU-R MIFR.

Notification and registration commenced in 1947 when the International Radio Conference created the International Frequency Registration Board (IFRB) to regulate the use of spectrum. The IFRB created a master frequency list where countries can notify new frequency registrations. The IFRB was replaced by the Radio Regulations Board (RRB) in 1995 (ITU, 2020). In order for administrations to register their assignments in the MIFR, they notify these assignments and then the ITU-R BR examines them with regard to their conformity with the RR. If accepted into MIFR, these assignments have international protection and recognition against harmful interference (Lyall, 2011). The main guiding principle of registering assignments in the MIFR is not to cause interference to assignments already made in accordance with the RR (ECC, 2013). Lyall (2011) argues that the ITU-R BR should be empowered to refuse registrations where there is an abuse of the process. Such abuse occurs where the MIFR accommodates some sort of declaration of spectrum use rather than actual use which creates apparent scarcity of spectrum (Struzak, 2007).

An important action that was initiated by the ITU-R BR occurred in May 2009 when the circular letter (CR/301) was issued urging administrations to remove unused frequency assignments and networks from the MIFR (ITU-R, 2009). The letter noted that there were some unused frequencies and GSO resources recorded in the MIFR that reduce the availability of opportunities to develop and launch new satellite networks. The letter also mentioned that similar cases had been examined by the RRB (ITU-R, 2009). The follow up to the letter resulted in the removal of 45% of the investigated cases and revealed that administrations did not have an incentive to report suspended networks (Man, 2016).

It is worth mentioning that in the early versions of the RR all services had the same priority but that in 1959 service priority was formally introduced in the format of primary, permitted and secondary services. Permitted services have a lower priority than primary services in the case of the preparation of frequency plans. In 1996, permitted services were removed (Ard-Paru, 2013). Secondary services are an older concept. The WRC-1932 approved the general radiocommunication regulations which stated that *'A fixed station may as a secondary service transmit to mobile stations on its normal working frequency subject to the following conditions: (a) the Administrations concerned consider it necessary to use this exceptional method of working; (b) no increase of interference results'* (ITU-R, 1932). The concept of secondary service was used in WRC-1938 when the USSR stated that they were to use broadcasting service as a secondary and supplementary service to the basic allocation of aeronautical services in certain spectrum bands (ITU-R, 1938). In addition, starting from WRC-1938, differentiation was associated with some services by indicating priority for them. Moreover, WRC-1947, added that *'a land station may communicate, on a secondary basis, with fixed stations or other land stations of the same category'* (ITU-R, 1947). The current form of primary and secondary services appeared relatively recently at WRC-1959, when the definition of primary and secondary services were stated explicitly (ITU-R, 1959).

In general, there are two types of services in terms of service allocation: planned (a priori) and non-planned services. The first type focuses on guaranteeing equitable access to orbit/spectrum resources for future use. Examples of these planned services include the allotment plan for the FSS using part of the 4/6 and 10–11/12–13 GHz frequency band and the plan for the BSS in the 11.7–12.7 GHz frequency band and the associated plan for feeder links in the 14 GHz and 17 GHz frequency bands. In contrast, non-planned services are based on first-come first-served coordination procedures with the aim of efficiency of orbit/spectrum use and interference-free operation satisfying actual requirements (ITU-R, 2020d).

4.3 Radiocommunication Service Flexibility

As we have earlier explained, the harmonization of services allocations has been the traditional approach of the ITU-R to handling interference. However, this has been criticized, and there has been a call for greater flexibility in the use of spectrum. The main argument in favour of flexibility is that service harmonization could lead to restrictions in the use of under- or unused spectrum for alternative uses and on the ability to refarm spectrum for new services (Chaduc & Pogorel, 2008; Indepen & Aegis Systems, 2004). Therefore, Cave (2002) advocates that harmonization should be time-limited until it enables manufacturers and operators to deliver a cost-effective service. Once this is achieved, the market should be opened up to other services. Harmonization can also impose constraints on spectrum use and add delays to the adoption of new services and technologies (Lie, 2004).

On the other hand, spectrum liberalization encourages innovation in new radio technologies and increases competition, but it may also reduce compatibility and interoperability and decrease the value of harmonized spectrum (Friederichs &

Mohyeldin, 2007). Spectrum liberalization also has implications for spectrum pricing and license conditions, where harmonization enables the regulator to levy different fees and conditions for the different services (Singh & Raja, 2008).

The harmonization of service allocation has been in place for several decades but the convergence among the different radiocommunication services (e.g. broadcasting, mobile, fixed) is relatively recent (El-Moghazi et al., 2012). Convergence in service allocation entails that differences between previously standalone services become increasingly blurred (e.g. unidirectional fixed point-to-multipoint system and broadcasting system) (Louis & Mallalieu, 2007). For instance, Wi-fi equipment operating in the 5 GHz are categorized under the mobile service in the RR, while they provide fixed broadband access (Radiocommunication Bureau, 2007). Furthermore, advancements in mobile technologies have enabled them to provide similar services to broadcasting such Multimedia Broadcast Multicast Service (eMBMS) in LTE (Watson, 2013). Together, this called for studying the alternatives for service allocation harmonization in order to bring more flexibility (Louis, 2011).

One alternative to the service allocation framework is to establish general service neutral values for interference parameters across all frequencies at the boundaries of countries and then modify these values through bilateral negotiations between countries (Lie, 2004). Another suggestion for service flexibility is to define generic service models based on anticipated use in addition to defining explicit transmit rights or explicit receive rights (Louis, 2011). This is in addition to the calls to view spectrum as an infrastructure asset used to deliver different applications to the end-user instead of the tradition approach of allocating spectrum exclusively to a particular service (Legutko, 2008). Moreover, a study by Eurostrategies and LS-Telecom (2007) outlines several spectrum management techniques that were split between users in different ways in additional to the traditional ITU way of splitting the radio spectrum by the type of service. Through this approach, interference is managed through the harmonization of frequency allocations into various generic radiocommunication services.

In practice, flexibility in service allocation has not been widely nor successfully adopted due to difficulties in its implementation (El-Moghazi et al., 2014). This is despite the argument that in those few cases where spectrum liberalization has been introduced, there has been a positive effect in terms of increasing the supply of spectrum to operators (Hazlett & Leo, 2010). However, in most cases, spectrum use has not changed, and trading entails only ownership change (London Economics, 2008). One of the few examples where spectrum use has changed occurred in the USA where Nextel's Specialized Mobile Radio (SMR) operations resulted in interference with operating networks (London Economics, 2008). Another example of the problems associated with flexibility also occurred in the USA where the FCC auctioned spectrum for a multipurpose 'wireless communications service' (WCS) which generated limited revenue due to the uncertainty related to interference (Goodman, 2004).

Not adopting flexibility in service allocation is related to the difficulties associated with this. Applying flexible spectrum use that is not in conformity with the ITU regulations would require extensive coordination with neighbouring countries (Indepen, 2001). This is dependent on the geographical characteristics of the countries in terms of area, coastal line length and number of neighbouring countries (Cave, 2002). Furthermore, service neutrality implies system architecture and duplexing modes neutrality (e.g. mesh network, point–multipoint, TDD, FDD) which is difficult to achieve in practice (Frullone, 2007).

Another reason for abandoning flexibility is related to the benefits of harmonization. Harmonization is critical for services of an international nature such as maritime and aeronautical purposes (Cave, 2005) and is also important for the industry because the number of spectrum bands supported by devices is limited (Ercole, 2009). Without harmonization, devices would be more complex and less sensitive which implies more base stations and additional costs (Christmas, 2007). Moreover, service harmonization has the benefits of facilitating roaming, enabling more standardized technologies and allowing for the evolution of existing technologies, but it may also limit spectrum trading and hoard valuable spectrum (Friederichs & Mohyeldin, 2007). Jones (1968) argued that international regulations do not restrict the transfer of spectrum rights as most domestic uses of the spectrum do not produce interference beyond the borders of the country. Freyens (2010) highlighted the importance of global harmonization provided through the ITU-R RR and explained that only geographically isolated countries that are simultaneously large equipment manufacturers can afford the costs associated with ignoring ITU regulations.

4.4 Elements of Flexibility Within the RR

Countries are sovereign, and they may decide to deviate from the RR allocation when it is difficult to relocate existing allocations or if they are bounded by multilateral agreements that are not reflected in the RR (Bonin et al., 2013). Nevertheless, flexibility is included in the RR with respect to service allocation via a number of different measures (El-Moghazi et al., 2017). Elements of flexibility include the ability of countries to deviate from the RR on the condition that it does not cause interference to neighbouring countries. Therefore, restrictions occur only in border areas and is dependent on the geographical position of the country. Moreover, the three categories of service allocation—primary and secondary and non-interference non-protection use—provide adequate flexibility according to RR Article 4.4. When combined with the RR, this enables having multiple service allocations in the same spectrum band, it is possible to introduce flexibility by having additional or alternative service allocations.

The RR also include several examples where there are forms of spectrum property rights. For example, stations operating in the 3400–3600 MHz band as a primary mobile service have to satisfy technical conditions in terms of power flux density at the border of the territory of any other administration (ITU-R, 2012c). Additionally, convergence at the application level does not necessarily mean convergence in radiocommunication service allocations. Thus, flexibility is already provided for in the current ITU-R service structure as the different applications could be provided through different types of radiocommunication services. Furthermore, the three ITU regional system provides flexibility to countries within each region to reach agreements independently from the other regions. Of course, such an approach could also result in those countries located on the borders of regions encountering limitations regarding their flexibility options.

One route to provide flexibility is the footnotes which allow additional or alternative allocations in specific cases. For instance, Footnote 5.237 provides for additional allocations in several countries in the 174–223 MHz band to fixed and mobile services on a secondary basis. Footnote 5.112 also provides an alternative allocation in the 2194–2300 kHz band to fixed and mobile services (except aeronautical mobile) on a primary basis in a few countries (ITU-R, 2020b). Footnotes can also indicate the priority of services, as well as a specific operation constraints (Ard-Paru, 2013).

Resolution 26 regulates the use of footnotes, demonstrating in the process that footnotes are an integral part of the RR (ITU-R, 2020e). The resolution resolves that *'wherever possible, footnotes to the Table of Frequency Allocations should confined to altering, limiting or otherwise changing the relevant allocations rather than dealing with the operation of stations, assignment of frequencies or other matters'*. It also highlights the roles of new footnotes which include providing flexibility, protect existing allocations and introduce restrictions on a new service to achieve compatibility. The resolution clarifies cases where the addition of a new footnote or modification of an existing footnote should be considered by a WRC or when the addition or modification of footnotes is specifically included in the agenda of the conference as a result of proposals being submitted by one or more interested administration.

There are, however, some reservations regarding the use of footnotes including their use being a threat to achieving global service allocation (Woolley, 1993). Furthermore, deleting a country's name from the footnotes is not an easy task as it may have implications on other countries included in the same footnote or whenever the footnote indicates an alternative allocation (ITU, 1997). In addition, while proposals to add country names to certain footnotes should be related to the agenda of WRC, every conference is sovereign so it can decide on its own course of action (ITU, 1998).

El-Moghazi et al. (2017) highlight the elements of flexibility provided by footnotes via additional or alternative service allocation or exemption from a service. Footnotes are considered as a starting point for harmonized allocation in the RR, as was in the case of the footnote indicating mobile primary allocation in the C-Band in Region 1 at the RR of 2008 (ITU-R, 2008). Such a footnote was useful for the mobile community as it accommodated a large number of countries which was, in effect, similar to a regional allocation in the RR. However, footnotes can also reduce flexibility where they provide priority to the incumbent service or remove an allocation from the RR. Moreover, the flexibility provided to some countries may reduce the flexibility to another country.

Another measure for providing flexibility is through special agreements as mentioned in Article 6.1 *'Two or more Member States may, under the provisions for special arrangements in the Constitution, conclude special agreements regarding the sub-allocation of bands of frequencies to the appropriate services of the participating countries'* (ITU-R, 2020c). An example of a special arrangement is Footnote

5.188 *'Additional allocation: in Australia, the band 85–87 MHz is also allocated to the broadcasting service on a primary basis. The introduction of the broadcasting service in Australia is subject to special agreements between the administrations concerned'* (ITU-R, 2020b). In this footnote, the 75.4–87 MHz band is allocated to fixed and mobile services in Region 3; however, Australia allocates the 85–87 MHz band for broadcasting service on the condition of having agreements with countries that may be affected by such an allocation.

An important contribution was made by Sweden in the context of Agenda Item. 1.1 of WRC-15, clarifying that service allocation should be differentiated from actual implementation and usage. This explains why, in several bands, there are allocations to more than one radiocommunication service. The contribution further suggests that the result of sharing and compatibility studies should be undertaken from the perspective of possible cross-border interference and not compatibility within a given country (Sweden, 2014). In other words, sharing studies should not address cosharing between different services (e.g. mobile and broadcasting) within the same country but rather among neighbouring countries. This should provide flexibility to countries when using the spectrum within their territories.

El-Moghazi et al. (2017) also investigate the flexibility provided within MIFR in terms of service allocation and highlight two views. The first is that MIFR usage is related to the international recognition of the spectrum used in a country with respect to other countries, and changing the assignments could be conducted at any time unlike the case of footnotes, which can only be amended by WRC. On the other hand, if the national assignments are flexible enough they will not be registered in the MIFR, which has a specific rule so that it would be possible to register the assignment under only one allocation. Therefore, there is a trade-off between flexibility and protection in cases where there is a need for flexibility, international protection would not be provided by the MIFR.

4.5 Flexibility in a Priori Planning

One approach to ensuring equitable access to spectrum internationally is 'a priori planning' or 'planned service' allocation which guarantees access to spectrum when each country submits its requirements at a world or regional planning conference. Such an approach emerged in 1919 with aeronautical services to promote carrier safety (Zacher, 1996). Following this, the importance of regional planning grew with the increasing using of broadcasting stations to prevent interference among them (Zacher, 1996). By the time of WRC-47, there were suggestions from the USA and other countries to conduct a priori planning for the entire spectrum in a single step, while the Soviet Union and other countries called for a gradual planning approach (Savage, 1989). The proposals, however, faded away due to the fear of enabling the dominance of developing countries in the ITU-R (Zacher, 1996). More specifically, a priori planning of the entire spectrum at that time would have enabled developing countries to guarantee their share of frequencies which would have been larger than the developed countries due to their numerical advantage.

It is important to understand that the concept of a priori planning was developed for planning services that needed such an approach—e.g. broadcasting and maritime—and it faces those problems associated with first-come first-served or non-planned approach (El-Moghazi et al., 2017). These difficulties are quite apparent in the case of satellite services, where countries can submit their requirements for orbital sites on a first-come first-served basis. A well-known illustration of the misuse of the non-planned approach was when Tonga submitted an excessive number of fillings for geostationary orbital sites over the Pacific Ocean between 1988 and 1990. The practice continued in what is sometimes called a 'paper satellites' strategy, where countries file several applications in an attempt to increase their chances in acquiring access to satellite slots. To address the impact of such a strategy, a time limit was placed on the launching of satellites in these orbital slots. If the satellites were not launched within the specified window, the allocation was voided (Ryan, 2012a). This illustrates the importance for services such as terrestrial broadcasting to adopt a priori planning approach to provide equal opportunity for all countries to have an adequate number of TV channels instead of relying on first-come first-served approach that may favour developed countries.

The main plan for analogue broadcasting in Region 1 was the Stockholm Plan (ST61), which governed the usage of spectrum for terrestrial television in Europe. The plan that covered Africa and the Arab countries was laid out in 1989 in Geneva under the name of GE-89. The GE-06 plan replaced both of these plans and accommodated a frequency plan for DVB-T and T-DAB in the VHF and UHF frequencies (174–230 MHz and 470–862 MHz) (Beutler, 2008). The plan was drafted by the ITU Regional Radio Conference held in 2006 (RRC-06), which planned the process of the switchover of terrestrial broadcasting services for radio and television from analogue to digital in Europe, Africa and the Middle East (Irion, 2009). It is worth mentioning that the GE-06 plan was initially developed to cover Europe and then African countries and their neighbouring countries wanted to join (Beutler, 2008). The GE-06 plan determined the end of the transition period of the analogue broadcasting services in the UHF band to be 17 June 2015 (GSMA, 2012).

Shortly following RRC-06, WRC-07 approved an additional allocation in the 790–862 MHz band to mobile service effectively from 17 June 2015 (ITU-R, 2007a). WRC-12 added an additional mobile allocation and IMT identification in the 698–790 MHz band in Region 1 (ITU-R, 2012i). That was in response to a request from ASMG and ATU during the conference and significant resistance from CEPT countries (El-Moghazi et al., 2015). Following WRC-12, 47 sub-Saharan African countries agreed on frequency coordination for a 2015 digital switchover so that these countries can allocate the 700 MHz and 800 MHz bands for mobile service (Standeford, 2013).

During WRC-15, the USA proposed adding an allocation to the mobile services and identification for IMT in the range 470–694/698 MHz except for the 608–614 MHz band in Region 2. The USA also proposed the mandatory application of RR Article 9.21, which would require explicit coordination agreements for the implementation of mobile service (USA, 2015). In Region 1, four Arab countries made a proposal at the beginning of the conference of having a footnote to identify the band for IMT in the 470–698 MHz band in those countries wishing to, given that the identification is subject to (1) GE-06 agreement with respect to the protection of the broadcasting service and (2) RR Article 9.21 with respect to the other primary services operating in the band (Egypt et al., 2015).

On the other hand, CEPT countries opposed the mobile allocation in the 470–694 MHz band as it was extensively used by broadcasting services and also for SAB/SAP applications. In addition, it was argued that the mobile allocation would constrain the future development of broadcasting services and that large separation distances are needed between stations of mobile and broadcasting services (CEPT, 2015). ASMG, RCC and ATU expressed similar arguments to CEPT (ASMG, 2015; ATU, 2015; RCC, 2015).

Eventually, it was agreed to have a new Agenda Item (A.I.) for WRC-23 to review the spectrum use and needs of existing services in the 470–960 MHz band in Region 1 (ITU-R, 2015e). In particular, ITU-R would, following WRC-19, conduct sharing and compatibility studies in the 470–694 MHz band in Region 1 between broadcasting and mobile services (ITU-R, 2015d). In Region 2, only the Bahamas, Barbados, Canada, the USA and Mexico were able to obtain identification in the 470–698 MHz frequency band subject to the RR Article 9.21 with regard to the broadcasting service where Belize and Columbia achieved identification in only the 614–698 MHz band. In Region 3, a few island countries—Micronesia, the Solomon Islands, Tuvalu and Vanuatu—obtained identification in the 470–698 MHz band, while Bangladesh, Maldives and New Zealand succeeded in identifying the 610–698 MHz band (ITU-R, 2015a). Meanwhile, many countries failed to add their names into the footnote as their neighbours objected (e.g. Pakistan, Papua New Guinea, India) and many countries welcomed the footnote on the condition that their neighbours did not add their names to it (ITU-R, 2015c).

It appears that the additional mobile allocations in the 600 MHz, 700 MHz and 800 MHz bands by WRC-15, WRC-12 and WRC-07, respectively, have raised concerns regarding the viability of a priori planning, especially for services such as terrestrial broadcasting. In particular, where there have been several plans for the broadcasting service in the past, it seems that the life cycle of these types of plans is becoming shorter due to the fast pace of technology advancements.

Although the main advantage of a priori planning is that it guarantees equitable access to spectrum, valuable spectrum is pre-allocated and may not be in use. For instance, for broadcasting service in the UHF band, a country may have the 470–862 MHz band preplanned and reserved for the terrestrial broadcasting service even though it does not have the required the number of TV channels to occupy the band. In such a case, unused spectrum is reserved for the broadcasting service in that country. Even Coase, who argued for spectrum flexibility several decades ago, criticized the US approach at WARC-1947 to preplan all spectra in a way similar to the current a priori planning as it restrict the flexibility of spectrum use (Coase, 1959). Therefore, it is useful to examine whether a priori planning accommodates flexibility to implement other radiocommunication services (e.g. mobile) rather than just those in the plan (e.g. broadcasting).

The main element of flexibility within GE-06 is the concept of allotment, where there is no specific location for transmitters within a particular service area (Foster, 2008). Allotment is defined in Article 1.17 of the RR as '*Entry of a designated frequency channel in an agreed plan, adopted by a competent conference, for use by one or more administrations for a terrestrial or space radiocommunication service in one or more identified countries or geographical areas and under specified conditions*'. Furthermore, a declaration was agreed among many countries that signed GE-06 that allows other uses rather than broadcasting on non-protection non-interference basis (Foster, 2008).

Another element of flexibility is the 'envelope concept' included in the GE-06 plan that allows the operation of analogue services beyond the end of the transition period. That was due to the disagreement between European and other countries over the date when analogue stations should cease, with the former wanting this to happen as soon as possible. Therefore, the envelope concept allows for the operation of analogue stations through a digital plan entry on the condition that it does not produce more interference nor claim protection. Within this concept, any entry in the plan can be used for alternative applications as long as the implementation does not exceed the interference envelope derived from the characteristics of the digital plan entry (ITU, 2006).

4.6 Proposals of Flexibility at WRCs

One of the first attempts to bring more flexibility into the RR, according to the best of our knowledge, was at the Nice Plenipotentiary Conference in 1989 when it was noted that the RR were complex, unwieldy and difficult to apply. Accordingly, a Voluntary Group of Experts (VGE) was established to study the simplification of the RR in order to enable flexibility in the use of radio spectrum (Woolley, 1995).

The VGE's recommendations were discussed at the WRC-95 and drafted in WRC Recommendations 34 (CEPT, 2007). It was recommended that future WRCs should, where possible, allocate frequency bands to the most broadly defined services with a view to providing the maximum flexibility to ITU Member States in spectrum use (ITU-R, 1995c). In addition, it was suggested that frequency bands should, if possible, be allocated on a worldwide basis. It has been argued that this first wave to revise the spectrum allocation principles was part of a bigger move towards ITU reform and restructuring to adapt to the changing telecommunications environment (Woolley, 1995).

The second wave came during WRC-03, motivated by the convergence in radiocommunications services as exemplified by services such as mobile TV. More specifically, radio systems were perceived to have downstream and upstream channels that could operate within the same or different radio services (ITU-R, 2003a). The issue was discussed under AI 1.21, which was used to consider progress of the ITU-R studies concerning the technical and regulatory requirements of terrestrial wireless interactive multimedia applications in accordance with Resolution 737 (WRC-2000), with a view to facilitating global harmonization (ITU-R, 2000b).

Resolution 737 notes the historical practice of frequency segmentation, with differences occurring between regions and services, in the Table of Frequency Allocations. It also mentions Recommendation 34 (WRC-95), derived from the recommendations of the VGE to study alternative allocation methods, merging of services, etc. The resolution also invited the ITU-R to pursue studies to develop common worldwide allocations and to identify spectrum suitable for new terrestrial wireless interactive multimedia technologies and applications (ITU-R, 2000a). The resolution also encouraged the review of approaches to worldwide spectrum identification to facilitate the harmonization of emerging terrestrial wireless interactive multimedia systems for the instant and flexible implementation of universal personal services.

During WRC-03, EU countries called for the development of measures to facilitate the use of common, worldwide frequency bands for implementation of terrestrial wireless interactive multimedia applications (CEPT, 2003). On the other hand, several countries countered this call arguing that no regulatory impediments have been identified, and therefore, there was no need to study the issue at the next WRCs (ASMG, 2003; USA, 2003).

Ultimately, Resolution 737 was 'suppressed'—which means, in practice, that it was removed—due to a lack of interest in the continuation of the studies. It was perceived that the existing regulatory framework has sufficient flexibility to minimize any regulatory constraints resulting from the existing definitions of services and coordination procedures (Gavrilov, 2003; United States Department of State, 2003). Having said this, reflecting technological developments, WRC-03 also decided to continue studying the effectiveness, appropriateness and impact of the radio regulations (ITU-R, 2003b). More specifically, WRC-03 resolved in Resolution 951 that the ITU-R would study the impact of the RR regarding the evolution of existing and future applications. It also instructed the Director of the BR to include the results of these studies in his report to WRC-07. Resolution 951 also considered that segregating bands for different radiocommunication services may not achieve the best possible spectrum efficiency and that the same technology could be used in different radiocommunication services and the same applications could combine different elements of radiocommunication services (ITU-R, 2003c).

Following WRC-03, studies were conducted within Study Group 1 and examples found of the convergence between different radiocommunication services. For instance, mobile radio devices with an ability to operate as receiving stations in the broadcasting service are an example of applications with elements of different radiocommunication services. DVB-H (Digital Video Broadcasting—Handheld) incorporates broadcasting service into mobile terminals (ITU-R, 2017). Moreover, in some cases, FSS networks are technically indistinguishable from BSS networks. The studies acknowledged that while WRC decision-making procedures may take too long, this was due to the complexity of the issues and the need to conduct sharing studies. Moreover, shortening the study period would decrease the flexibility of decision-making processes. It was also recognized that countries may adopt the flexibility within Article 4.4 to operate on a non-interference non-protection basis such as in the case of RLAN operation in the 2.4 GHz in the ISM bands.

Regarding the definitions of radiocommunication services, the studies acknowledged convergence. Convergence has occurred at the service and technology levels and in the core IP networks where the concurrent provision of fixed, mobile and broadcasting applications is possible. It was highlighted that differentiation between the mobile and fixed services has become more difficult as there have been applications that could be considered as part of either the fixed or mobile service (e.g. Electronic News Gathering). This has caused difficulties, especially with respect to notifying stations in the MIFR.

Eventually, the studies suggested four options to improve the international spectrum regulatory framework. The first option was to keep the current practice as there was sufficient flexibility within the present RR and WRC processes. In cases where a shorter timeframe is needed, the operation with no protection against interference according to Article 4.4 of the RR is possible. Furthermore, existing services definitions have been able to meet the requirements of emerging wireless applications. The second option was to review and possibly revise some of the current service definitions. This option could include adding a new service to the list of service definitions, which would encompass several of the existing ones. The third option was to introduce a new provision into the RR enabling substitution between assignments of specific services. For instance, assignments recorded in the MIFR for fixed service could be also used for mobile service provided that such transmissions do not cause more interference, or require more protection from interference, than the corresponding assignments in the fixed service recorded in the MIFR.

The issue was discussed under A.I. 7.1 of WRC-07, which considered and approved the report of the Director of the BR on the activities of the Radiocommunication Sector since WRC-03. It was recognized that changing the current services definitions would be impractical and no consensus would be reached (Sims, 2007). The issue was not resolved at WRC-07, which decided to continue studying the issue until WRC-12 under AI 1.2 (ITU-R, 2007b).

Following WRC-07, the ITU-R studies approached AI 1.2 from two perspectives. The first focused only on the convergence between terrestrial (fixed and mobile) services while the second addressed spectrum allocation issues more generally (ITU-R, 2011b). In addition, Resolution 951 was amended to explicitly include four options for enhancing the international spectrum regulatory framework:

- Retaining current practices;
- Revising the current service definitions or adding new services;
- Enabling substitutions between the assignment of specific services; and
- Introducing composite services.

Regarding the second perspective, two methods were proposed, namely keeping the current practice and introducing no change to the RR and having a WRC Resolution complementing the existing provisions in the RR. Furthermore, the revised resolution included guidelines for implementing the resolution, identifying three steps in the process (ITU-R, 2007b).

Regarding the first perspective, the convergence between terrestrial (fixed and mobile) services, the studies found that there is a frequent joint allocation to fixed and mobile services that may not be sufficient. Regarding the convergence between space services (FSS, MSS and BSS), it was found that this should be addressed on a case-by-case basis to avoid additional burdens being placed on terrestrial services in the frequency and due to the coordination and notification procedures differences among these services (ITU-R, 2011b). Finally, the studies showed that the number of worldwide allocations is 12.8% of total frequency band allocations and that only 41.73% of the spectrum is allocated to broadly defined services (ITU-R, 2011a). In other words, the need for immediate action is less pressing than initially expected.

During WRC-12, the APT, ASMG and RCC called for retaining current practices as there was sufficient flexibility within the current regulatory framework and the WRC process not to impede the introduction of new technologies. (APT, 2011; ASMG, 2011; RCC, 2011). In addition, the CITEL group proposed to change the definitions of fixed service and fixed and mobile stations (CITEL, 2012). The CEPT group also proposed to upgrade Recommendation 34 so that it would become more binding on countries (CEPT, 2011). More specifically, one proposal was to have a WRC resolution so that, whenever possible, spectrum is allocated to the most broadly defined service to achieve the maximum possible flexibility on a worldwide basis (ITU-R, 2011a). With regard to convergence between fixed and mobile services, European countries and a few others proposed to change some of the definitions within the RR (CEPT, 2011; CITEL, 2012). In general, all the regional organizations called for considering the work that has been completed regarding how to enhance the international spectrum regulatory framework.

Eventually, WRC-12 decided not to change the current spectrum allocation practices (ITU-R, 2012j) and to end the general studies that had been undertaken to enhance the international spectrum regulatory framework. They had been considered by two successive conferences without coming to a resolution (ITU-R, 2012f). Nevertheless, it was decided to continue with those studies intent on revising the definitions of fixed service, fixed station and mobile station until the next WRC in 2015 (ITU-R, 2012h). WRC-12 also considered that the current technological environment for some applications was substantively different from the one which prevailed when the current definitions were established and thus sought a review of the definitions of fixed services, fixed stations and mobile stations to study the potential impact on regulatory procedures on the RR (ITU-R, 2012h).

WRC-12 considered in Recommendation 16 the principles behind radiocommunication services and spectrum allocation that had first been adopted in Berlin in 1906 so that countries can obtain flexibility without causing harmful interference (ITU-R, 2012g). In addition, WRC-12 recommended that ITU-R studies all aspects of interference management resulting from the impact of technical convergence on the radio regulatory environment, involving stations that may be involved in cross-border interference (ITU-R, 2012d).

The studies following WRC-12 showed that the definition of a station in the RR implies that it may not, in principle, operate more than one radiocommunication service at the same time (ITU-R, 2012e). Another highlight from the studies following WRC-12 is that interference is sometimes managed within the RR through what is called 'explicit transmit rights' and 'explicit receive right'. In the former, spectrum is managed through the specification of parameters such as maximum transmitted power, while in the latter, the focus is on the receiver side (ITU-R, 2012a). The studies also revealed the challenges associated with changing the definitions of current radiocommunication services such as its impact on coordination and registration of frequency assignments of the stations (ITU-R, 2012b).

Prior to WRC-15, the positions of the regional organization groups regarding changing radiocommunication service definitions did not alter. For instance, ASMG supported no change to the definitions since it would affect national regulatory instruments (ASMG, 2013). Moreover, RCC felt that current definitions in the RR do not prevent the use of existing applications in fixed and mobile services (RCC, 2014). APT argued that here is no need to modify the existing definitions of a fixed service or fixed and mobile stations (APT, 2013). Canada, which was one of the main proponents of change, felt that this issue needed to be resolved at WRC-15 (CITEL, 2013).

Surprisingly, while CEPT was the main promoter of changing some of the radiocommunication service definitions to accommodate convergence prior to WRC-12, this changed leading up to WRC-15. This aligned CEPT with the other regional organizations. CEPT stated that it was '*... of the view that there is no need to modify the existing definitions of fixed service, fixed station and mobile station. Furthermore CEPT opposes any modification which may have any negative regulatory impact on existing allocations to radiocommunication services*' (CEPT, 2013). By 2013, it was clear that there was a lack of interest in both revising the various definitions and addressing management interference issues (ITU-R, 2013). As a result, during WRC-15, it was evident that there is no intention to amend these long-standing definitions, and it was thus decided not to change the RR during the first week of the conference (2015b; ITU-R, 2014).

4.7 Wireless Access Policy for Electronic Communication Services

Wireless Access Policy for Electronic Communication Services (WAPECS) is a concept that was developed in Europe to enable greater flexibility within the European Union (EU) spectrum management framework by allowing the use of spectrum on a technology and service neutral basis within certain technical requirements that avoids interference (RSPG, 2005). WAPECS was the response to the different calls within the EU to reform the traditional 'command and control' approach to spectrum management towards a more market-driven approach (Gulyaev, 2011; Selek, 2008). The roots of these calls can be traced back to the European Commission (EC) green paper on spectrum policy that sought to facilitate competition via flexible planning of spectrum use (European Commission, 1998). Moreover, WAPECS was proposed within the context of the i2010 Information Society Initiative to ensure that spectrum is available across a wide variety of services and applications within the EU (Akalu, 2006).

WAPECS was also motivated by the perception that any communication service could be delivered through any platform and that removing restrictions associated with individual spectrum bands would promote competition between the different delivery systems (Forge et al., 2012). It was perceived that the liberalization of spectrum use in general and WAPECS more specifically could limit anti-competitive behaviour (Cave, 2010). There was also a call by 2G operators for greater flexibility in their licenses as they moved towards 3G and the need to take into account the impact of technological convergence (Delaere, 2007).

In 2004, the EC issued a request for opinions to the Radio Spectrum Policy Group (RSPG) to develop and adopt an opinion on a coordinated EU spectrum policy approach concerning WAPECS. The RSPG conducted a public consultation on WAPECS, with different views being received from industry participants (ComReg, 2005). There was limited agreement among those who replied as both broadcasters and mobile operators sought to protect their interests (Sutherland, 2006). Furthermore, most of the responses did not support the definition of WAPECS, arguing that it is not clear what it is intended to address. They also disagreed to the use of the term ‘Platform’, with ‘System’ being suggested instead. The RSPG issued their final opinion on WAPECS in 2005 when it was decided to change ‘Platform’ into ‘Policy’ (RSPG, 2005). Therefore, the acronym of WAPECS changed to ‘Wireless Access Policy for Electronic Communications Services’.

Within WAPECS, the traditional service and technology combination was replaced by what was called ‘electronic communications services’ where different applications (e.g. voice, data and video) could be provided through a variety of technologies and networks (Chaduc & Pogorel, 2008). The implementation of WAPECS assumes that the most likely use of the spectrum is by technologies using a cellular network topology based on a two-way communication system with a network of base stations and associated subscribers (CEPT, 2010).

In 2006, the EC instructed CEPT to develop less restrictive technical conditions in the frequency bands addressed in the context of WAPECS (European Commission, 2006). Accordingly, CEPT worked on the issue, and the Block Edge Mask (BEM) methodology was chosen to identify the least restrictive technical conditions for frequency bands with a focus on the 2.5 GHz and 3.5 GHz spectrum bands (European Commission, 2008a, 2008b). It is worth mentioning that CEPT developed similarly less restrictive technical conditions based on BEM for the 790–862 MHz and 2 GHz bands (CEPT, 2009, 2010). One of the major influences of WAPECS implementation was the ending of the exclusivity enjoyed by GSM in the 900 MHz band and thus allow the operation of other systems in it (Guijarro & Alabau, 2013).

WAPECS has been discussed, albeit briefly, in the literature. Firstly, Roetter (2009) criticized the decision of the EC in 2008 that determined that the 2570–2620 sub-band MHz could be used by TDD and that the usage in the rest of the band can be decided nationally to be TDD or FDD (European Commission, 2008a). In particular, Roetter (2009) explained that interference management would be quite difficult due to the need to coordinate between diverse arrangements of TDD and FDD spectrum blocks. Furthermore, such a decision increases the uncertainty among operators regarding how much usable bandwidth they will eventually acquire if they win the frequencies that they decide to bid for. In fact, the auctions that occurred across Europe up until 2010 showed that most countries favoured a fixed TDD-FDD plan for the band (Marsden et al., 2010). Forge et al. (2012) pointed out the paradox mentioned in one of the CEPT reports on the issue of flexibility: introducing more flexibility in spectrum use management restricted the flexibility of band use (CEPT, 2006). Furthermore, Forde et al. (2010) highlight that WAPECS shifts the consideration of specific technologies while shifting BEM towards consideration of different network scenarios instead.

Regarding the interaction between WAPECS and ITU-R radiocommunication services, Grad (2011) explained that WAPECS was based on the convergence between three main radiocommunication services (fixed, mobile and broadcasting). This was also confirmed by Akalu (2006), who pointed out that WAPECS aimed to enable these services to be provided over different access networks in order to compete with each other. Louis (2011) noted that the models proposed for sharing under WAPECS could be perceived as alternatives for radiocommunication service harmonization internationally. Moreover, Forge et al. (2012) highlight the constraint emerging from international agreements and the need to keep bands harmonized even if the aim is to introduce greater flexibility.

One of the main constraints in the application of WAPECS, according to RSPG, is the lack of flexibility in some existing licenses, particularly due to the international agreements that have been undertaken (RSPG, 2005). Nevertheless, the concept of service neutrality within the WAPECS concept is different from that of neutrality within the ITU-R radiocommunication services. The former is based on replacing traditional service and technology combinations by ‘electronic communications services’ (ECS) where mobile, portable or fixed access could be provided under one or more radiocommunication service allocations (Chaduc & Pogorel, 2008). This is shown in Fig. 4.2.

As shown in Fig. 4.2, it appears that there is intersection between ECS and the ITU-R radiocommunication services as WAPECS covers several radiocommunication services. It can, however, be argued that the RR can accommodate WAPECS except in the case of some services like satellite as it was designed to be deployed in bands where there are already coprimary service allocations. In addition, WAPECS was perceived as a national policy and that the RR are concerned with international issues between neighbouring countries (El-Moghazi et al., 2014).

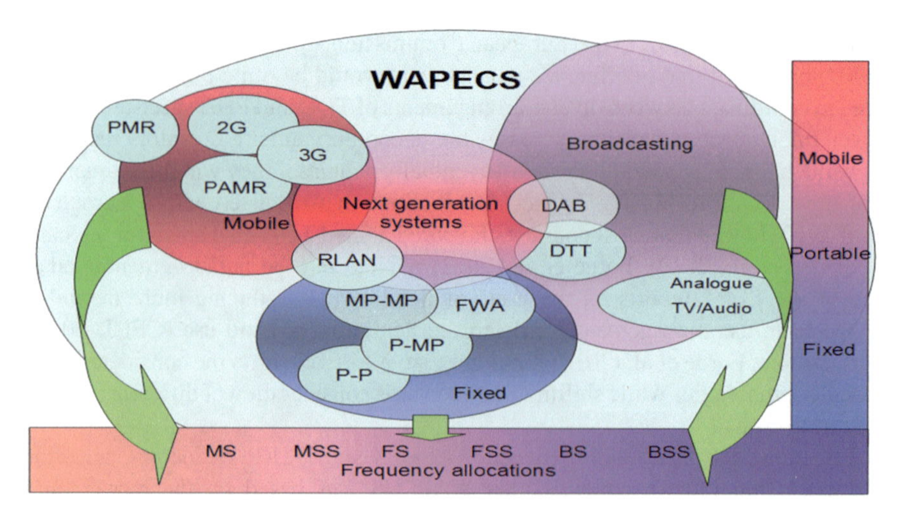

Fig. 4.2 Radiocommunication service allocation covered by WAPECS. *Source* RSPG (2005)

4.8 Summary

In this chapter, we have addressed radiocommunication service allocation flexibility where defining spectrum property rights with usage flexibility is a key issue in spectrum markets. One of the dimensions of flexibility is to have neutrality in radiocommunication service allocation where within the same bands several radiocommunication services could be deployed (e.g. fixed, mobile, broadcasting). Article 5 of the RR divides the frequency band from 8.3 kHz to 3000 GHz into smaller bands that are allocated to more than 40 different radiocommunication services. The ITU-R RR divides the world into three regions in terms of spectrum allocation, where each spectrum band could be allocated to one or more radio service with equal or different rights (primary and secondary).

Countries can register their frequency assignments via two different approaches. The first is called 'a priori' planning, which enables the guaranteeing of access to the spectrum with each country submitting its requirements at a global or regional planning conference. The second type of assignment registration within the non-planned bands is on a 'first come, first-served' basis in the MIFR. While services allocations harmonization has been the traditional approach of the ITU-R to handle interference, flexible allocation can encourage innovation in new radio technologies and increases competition. There has recently been convergence in service allocation which results in the previously standalone services become increasingly blurred; however, flexibility in service allocations has not been widely nor successfully adopted in practice due to the difficulties in its implementation.

One approach to ensuring equitable access to spectrum internationally is 'a priori planning' or 'planned service' allocation that guarantees access to spectrum when each country submits its requirements at a world or regional planning conference.

Although the main advantage of a priori planning is that it guarantees equitable access to spectrum, valuable spectrum is pre-allocated and may not be used. The main element of flexibility within GE-06, which is a practical example of a priori planning in broadcasting, is the concept of allotment, where there is no specific location for transmitters within a particular service area. Another element of flexibility is the 'envelope concept' included in the GE-06 plan that allows the operation of analogue services beyond the end of the transition period.

There have been several proposals to introduce greater flexibility into service allocation. It was recommended in 1995 that future WRCs should, where possible, allocate frequency bands to the most broadly defined services with a view to providing the maximum flexibility to ITU countries in spectrum use. The second wave came during WRC-03, motivated by the convergence in radiocommunications services as exemplified by services such as mobile TV.

Following WRC-03, studies were conducted within Study Group 1 and found examples of the convergence between different radiocommunication services. Following WRC-07, ITU-R studies approached the issue of flexibility from two perspectives. The first focused only on the convergence between terrestrial (fixed and mobile) services, while the second addressed spectrum allocation issues more generally.

Notwithstanding the studies that have been undertaken and the examples of convergence that have emerged, WRC-12 opted not to change how spectrum is allocated and resolved to review the definitions of fixed service, fixed station and mobile station. During WRC-15, it was evident that there is no intention to amend these long-standing definitions, with the consequence that it was decided not to change the RR.

While WAPECS demonstrated what could be achieved with regard to flexibility, it is worth remembering that it is viewed as being a nationally focused development and thus did not impact on the internationally orientated RR.

References

Akalu, R. (2006). EU spectrum reform and the wireless access policy for electronic communications services (WAPECS) Concept. *Info, 8*(6), 31–50.

APT. (2011). *Common Proposals for the Work of the Conference. Agenda Item 1.2* World Radiocommunication Conference (WRC-12), Geneva. Retrieved from https://www.itu.int

APT. (2013). *APT Preparatory Process for WRC-15* 1st ITU Inter-Regional Workshop on WRC-15 Preparation, Geneva. Retrieved from https://www.itu.int

Ard-Paru, N. (2012). *Information and coordination in international spectrum policy: Implications for Thailand*. Retrieved from https://research.chalmers.se

Ard-Paru, N. (2013). *Implementing spectrum commons: Implications for Thailand*. Chalmers University of Technology, PhD thesis.

Ascom. (2012). *TD-LTE and FDD-LTE: A Basic Compariosn.* http://www.ascom.com/

ASMG. (2003). *World Radiocommunications Conference 2003: Arab States Common proposal.* Retrieved from https://www.itu.int

ASMG. (2011). *Arab States Common Proposals. Common Proposals for the Work of the Conference. Agenda Item 1.2* World Radiocommunication Conference (WRC-12), Geneva. Retrieved from https://www.itu.int
ASMG. (2013). *Agenda Items of the WRC-15* 1st ITU Inter-Regional Workshop on WRC-15 Preparation, Geneva. Retrieved from https://www.itu.int
ASMG. (2015). *Arab States Common Proposals. Common Proposals for the Work of the Conference. Agenda Item 1.1* World Radiocommunication Conference (WRC-15), Geneva. Retrieved from https://www.itu.int
ATU. (2015). *African Common Proposals for the Work of the Conference. Agenda Item 1.1* World Radiocommunication Conference (WRC-15), Geneva. Retrieved from https://www.itu.int
Beutler, R. (2008). The regional radio communication conference RRC-06 and the GE06 agreement. In *Digital terrestrial broadcasting networks*. Springer.
Bonin, J.-P., Evc, C., & Sanders, A. L. (2013). Securing spectrum through the ITU to fuel the growth of next-generation wireless technologies. *Bell Labs Technical Journal, 18*(2), 99–115.
Cave, M. (2002). *Review of radio spectrum management, an independent review for department of trade and industry and HM treasury*. www.ofcom.org.uk
Cave, M. (2005). *Independent audit of spectrum holdings: Final report*. www.ofcom.org.uk
Cave, M. (2010). Anti-competitive behaviour in spectrum markets: Analysis and response. *Telecommunications Policy, 34*(5–6), 251–261.
CEPT. (2003). *WRC-03: European Common PROPOSALS FOR THE WORK OF THE Conference*. Retrieved from https://www.itu.int
CEPT. (2006). *CEPT report 80: Enhancing harmonisation and introducing flexibility in the spectrum regulatory framework*. Retrieved from http://www.erodocdb.dk
CEPT. (2007). *Draft CEPT Brief on Agenda Item 7.1 (Resolution 951 (WRC-03))*. Retrieved from http://www.anacom.pt
CEPT. (2009). *CEPT Report 30: Report from CEPT to the European Commission in Response to the Identification of Common and Minimal (least restrictive Technical Conditions for 790 - 862 MHz for the Digital Dividend in the European Union"*. Retrieved from http://www.erodocdb.dk
CEPT. (2010). *CEPT Report 39: Report from CEPT to the European Commission in Response to the Mandate to Develop Least Restrictive Technical Conditions for 2 GHz bands*. Retrieved from http://www.erodocdb.dk
CEPT. (2011). *European Common Proposals for the Work of the Conference, Part 1, Agenda Item 1.2* World Radiocommunication Conference (WRC-12), Geneva. Retrieved from https://www.itu.int
CEPT. (2013). *Status of CEPT Preparations for WRC-15* 1st ITU Inter-Regional Workshop on WRC-15 Preparation, Geneva. Retrieved from https://www.itu.int
CEPT. (2015). *European Common Proposals for the Work of the Conference, Agenda Item 1.1* World Radiocommunication Conference (WRC-15), Geneva. Retrieved from https://www.itu.int
Chaduc, J., & Pogorel, G. (2008). *The Radio Spectrum. Managing a Strategic Resource*. ISTE Ltd.
Christmas, F. (2007). *Benefits of Frequency Harmonisation* ITU Workshop on Market Mechanisms for Spectrum Management, Geneva.
CITEL. (2012). *Inter-American Proposals for the Work of the Conference. Agenda Item 1.2* World Radiocommunication Conference (WRC-12), Geneva. Retrieved from https://www.itu.int
CITEL. (2013). *Preliminary Views and Proposals Regarding WRC-15 Agenda Items* 1st ITU Inter-Regional Workshop on WRC-15 Preparation, Geneva. Retrieved from https://www.itu.int
Coase, R. H. (1959). The Federal Communications Commission. *Journal of Law & Economics, 2*(1), 1–40.
ComReg. (2005). *Summary of the Analysis of Responses to RSPG Consultation on WAPECS*. Retrieved from http://rspg-spectrum.eu/
Delaere, S. (2007). European policy trends towards flexible spectrum management. *The Southern African Journal of Information and Communication, 1*(8), 8–29.
ECC. (2013). *ECC Reports 205: Licensed Shared Access (LSA)* ECC Reports, Issue.

Egypt, Jordan, Lebanon, & Morocco. (2015). *World Radiocommunications Conference 2015: Proposals for the Work of the Conference. Agenda Item 1.1.* Retrieved from https://www.itu.int
El-Moghazi, M., Whalley, J., & Irvine, J. (2012). *WRC-12: Implication for the spectrum eco-system.* Telecommunications Policy Research Conference 2012, Arlington.
El-Moghazi, M., Whalley, J., & Irvine, J. (2014). *WAPECS: Collision between practice and theory.* ITS 2014 Conference, Brussels.
El-Moghazi, M., Whalley, J., & Irvine, J. (2015). *The 700 MHz mobile allocation in Africa: Observations from the Battlefield.* CPR Asia/CPR Africa, Taiwan.
El-Moghazi, M., Whalley, J., & Irvine, J. (2017). International spectrum management regime: is gridlock blocking flexible spectrum property rights? *Digital Policy, Regulation and Governance, 19*(2), 1–13.
Ercole, R. (2009). *Spectrum allocation: An industry perspective.* Retrieved from http://www.gsma.com/
European Commission. (1998). *COM (98) 596 Final: Green Paper on Radio Spectrum Policy.*
European Commission. (2006). *Mandate to CEPT to Develop Least Restrictive Technical Conditions in Frequency Bands Addressed in the Context of WAPECS.* Official Journal of the European Union.
European Commission. (2008a). Commission Decision of 13 June 2008 on the Harmonisation of the 2500–2690 MHz Frequency Band for Terrestrial Systems Capable of Providing Electronic Communications Services in the Community. *Official Journal of the European Union, L 163.*
European Commission. (2008b). Commission Decision of 21 May 2008 on the harmonisation of the 3 400–3 800 MHz frequency band for terrestrial systems capable of providing electronic communications services in the Community. *Official Journal of the European Union, L 144.*
Eurostrategies and LS-Telecom. (2007). *Study on Radio Interference Regulatory Models in the European Community.* Retrieved from https://op.europa.eu
Forde, T. K., Doyle, L. E., & Ozgul, B. (2010). *Dynamic Block-Edge Masks (BEMs) for Dynamic Spectrum Emission Masks (SEMs)* IEEE Symposium on New Frontiers in Dynamic Spectrum Access Networks, Singapore.
Forge, S., Horvitz, R., & Blackman, C. (2012). *Perspectives on the value of shared spectrum access: Final Report for the European Commission.* Retrieved from http://ec.europa.eu
Foster, A. (2008). *Spectrum Sharing.* GSR 2008, Thailand. Retrieved from https://www.itu.int
Foster, A., Cave, M., & Jones, R. W. (2011). Going mobile: Managing the spectrum. In C. Blackman & L. Srivastava (Eds.), *Telecommunications regulation handbook.* The World Bank.
Freyens, B. P. (2010). Shared or exclusive radio waves? A Dilemma gone astray. *Telematics and Informatics, 27*(3), 293–304.
Friederichs, K.-J., & Mohyeldin, E. (2007). *Cognitive radio impacts on spectrum management: Liberalisation and Harmonisation* Software Defined Radio (SDR)/Cognitive Radio (CR) Workshop Sophia.
Frullone, M. (2007). *A Deeper Insight in Technology and Service Neutrality* ITU Workshop on Market Mechanisms for Spectrum Management, Geneva.
Gavrilov, T. (2003). *Results of the WRC-03* Regional Radiocommunication Seminar, Lusaka. Retrieved from https://www.itu.int
Goodman, E. P. (2004). Spectrum Rights in the Telecosm to Come. *San Diego Law Review, 41*(269).
GSMA. (2012). *Geneva 06: Regional Radio Conference.* Retrieved 30/12 from http://www.gsma.com
Guijarro, L., & Alabau, A. (2013). *Radio spectrum and broadband European Union policies: Towards a European Broadband Strategy* CPR. Europe, Brussels.
Gulyaev, A. (2011). *WAPECS: Harmonisation vs flexibillity - Towards increasing the flexibility of spectrum use in Europe* CEPT Workshop 2011, Copenhagen.
Hazlett, T., & Leo, E. (2010). The case for liberal spectrum licenses: A technical and economic perspective. *George Mason Law & Economics, Research Paper No. 10-19.*
Indepen, A. (2001). *Implications of international regulation and technical considerations on market mechanisms in spectrum management: Report to the independent spectrum review.* http://www.ofcom.org.uk

Indepen and Aegis Systems. (2004). *Costs and benefits of relaxing international frequency harmonisation and radio standards*. stakeholders.ofcom.org.uk

Irion, K. (2009). Separated together: The international telecommunications union and civil society. *International Journal of Communications Law and Policy, 13.*

ITU. (1997). WRC-97 News. *WRC News*. Retrieved December 30, 2014, from https://www.itu.int

ITU. (1998). The challenges of spectrum allocation and Sharing: An interview with Veena Rawat. *ITU News* (1).

ITU. (2006). Annex 2: Technical Elements and Criteria Used in the Development of the Plan and the Implementation of the Agreement. In *Final Acts of the Regional Radiocommunication Conference for Planning of the Digital Terrestrial Broadcasting Service in Parts of Regions 1 and 3, in the Frequency Bands 174–230 MHz and 470–862 MHz (RRC-06)*. ITU.

ITU. (2020). *International Frequency Registration Board (IFRB), 1947–1993*. Retrieved September 25, 2020 from https://www.itu.int

ITU-R. (1932). General radiocommunication regulations. In *International telecommunication convention*. Retrieved from https://www.itu.int

ITU-R. (1938). General radiocommunication regulations. In *International telecommunication convention*. Retrieved from https://www.itu.int

ITU-R. (1947). General radiocommunication regulations. In *International telecommunication convention*. Retrieved from https://www.itu.int

ITU-R. (1959). General radiocommunication regulations. In *International telecommunication convention*. Retrieved from https://www.itu.int

ITU-R. (1995a). ITU-R Recommendation SM.1131 factors to consider in allocating spectrum on a worldwide basis. In *SM Series. Spectrum Management*. ITU.

ITU-R. (1995b). ITU-R Recommendation SM.1133: Spectrum utilization of broadly defined services. In *SM Series. Spectrum Management*. ITU.

ITU-R. (1995c). *WRC-95 Recommendation 34. Principles for the allocation of frequency bands*. ITU.

ITU-R. (2000a). Resolution 737 (WRC-2000): Review of spectrum and regulatory requirements to facilitate worldwide harmonization of emerging terrestrial wireless interactive multimedia applications. In *WRC-2000 Final Act*. ITU.

ITU-R. (2000b). *Resolution 800 (WRC-2000): Agenda for the 2003 World Radiocommunication Conference* Final Acts: WRC-2000, Istanbul.

ITU-R. (2003a). *CPM Report on Technical, Operational and Regulatory/Procedural Matters to be Considered by the 2003 World Radiocommunication Conference*. ITU.

ITU-R. (2003b). Resolution 951: Options to Improve the International Spectrum Regulatory Framework. In *Provisional Final Acts - World Radiocommunication Conference (WRC-03)*. ITU.

ITU-R. (2003c). *Resolution 951: Options to improve the international spectrum regulatory framework*. RR 2004. ITU.

ITU-R. (2007a). *Article 5: Frequency allocations*. RR 2008. ITU.

ITU-R. (2007b). WRC-07 Resolution 951: Options to improve the international spectrum regulatory framework. In *Provisional Final Acts - World Radiocommunication Conference (WRC-07)*. ITU.

ITU-R. (2008). Article 5: Frequency allocations. In *Radio Regulations*. ITU.

ITU-R. (2009). *CR/301: Removal of unused frequency assignments (Space Services) from the Master Register*. Retrieved from https://www.itu.int

ITU-R. (2011a). *Annex 5 to Working Party 1B Chairman's Report: Working Document Towards a Preliminary New Report ITU-R SM. [RES. 951]*. Retrieved from https://www.itu.int

ITU-R. (2011b). *CPM Report on Technical, Operational and Regulatory/Procedural Matters to be Considered by the 2012 World Radiocommunication Conference*. ITU.

ITU-R. (2012a). *Annex 3 to Report of the Meeting of Working Party 1B: Working Document towards A Draft New Report ITU-R SM.[INTERF_MNGNT]*. Retrieved from https://www.itu.int

ITU-R. (2012b). *Annex 11 to Working Party 1B Chairman's Report WRC-15 Agenda item 9.1, Issue 9.1.6 (Resolution 957 (WRC-12)) Studies towards Review of the Definitions of Fixed service, Fixed Station and Mobile Station*. Retrieved from https://www.itu.int

ITU-R. (2012c). Article 5: Frequency allocations. In *Radio Regulations*. ITU.
ITU-R. (2012d). *Recommendation COM6/2: Interference Management for Stations that May Operate Under More than One Terrestrial Radiocommunication Service* (WRC-12 Final Acts, Issue. Retrieved from https://www.itu.int
ITU-R. (2012e). *Report of the Meeting of Working Party 1B*. Retrieved from https://www.itu.int
ITU-R. (2012f). *Resolution 804: Principles for Establishing Agendas for World Radiocommunication Conferences*. ITU.
ITU-R. (2012g). WRC-12 Recommendation 16: Interference Management for Stations that May Operate Under More than One Terrestrial Radiocommunication Service. In *Provisional Final Acts - World Radiocommunication Conference (WRC-12)*. ITU.
ITU-R. (2012h). WRC-12 Resolution 957: Studies Towards Review of the Definitions of Fixed service, Fixed Station and Mobile Station. In *Provisional Final Acts - World Radiocommunication Conference (WRC-12)*. ITU, Geneva.
ITU-R. (2012i). WRC-12 Resolution 232. Use of the Frequency 694–790 MHz by the Mobile, Except Aeronautical Mobile, Service in Region 1 and Related Studies. In *Provisional Final Acts - World Radiocommunication Conference (WRC-12)*. ITU.
ITU-R. (2012j). WRC-12 Weekly Highlights. (4). Retrieved March 19, 2012, from https://www.itu.int
ITU-R. (2013). Annex 6 to Working Party 1B Chairman's Report: Working Document towards Draft CPM text on WRC-15 Agenda Item 9.1, Issue 9.1.6: Resolution 957 (WRC-12) – Studies towards Review of the Definitions of Fixed Service, Fixed Station and Mobile Station. In *Working Party 1B Chairman's Report*.
ITU-R. (2014). *CPM Report on Technical, Operational and Regulatory/Procedural Matters to be Considered by the 2015 World Radiocommunication Conference*. Retrieved from https://www.itu.int
ITU-R. (2015a). Article 5: Frequency Allocations. In *Radio Regulations*. ITU.
ITU-R. (2015b). *Provisional Final Acts - World Radiocommunication Conference (WRC-15)*. ITU.
ITU-R. (2015c). *Summary Record of the Twelfth Plenary Meeting* World Radiocommunication Conference (WRC-15), Retrieved from https://www.itu.int
ITU-R. (2015d). WRC-15 Resolution COM4/6 (WRC-15), Review of the spectrum use of the frequency band 470–960 MHz in Region 1. In *Provisional Final Acts - World Radiocommunication Conference (WRC-15)*. ITU.
ITU-R. (2015e). WRC-15 Resolution COM 6/2: Agenda for the 2023 World Radiocommunication Conference. In *Provisional Final Acts - World Radiocommunication Conference (WRC-15)*. ITU.
ITU-R. (2017). Report ITU-R BT.2295–2: Digital terrestrial broadcasting system. In *ITU-R Reports BT Series*. ITU.
ITU-R. (2020a). Article 1: Terms and definitions. In *Radio Regulations*. ITU.
ITU-R. (2020b). Article 5: Frequency allocations. In *Radio Regulations*. ITU.
ITU-R. (2020c). Article 6: Special agreements. In Radio Regulations. ITU.
ITU-R. (2020d). *Frequently Asked Questions (FAQ) related to Space Plans*. Retrieved September 22, 2020 from https://www.itu.int
ITU-R. (2020e). Resolution 26: (Rev.WRC-19) Footnotes to the Table of Frequency Allocations in Article 5 of the Radio Regulations. In *Radio Regulations*. ITU, Geneva.
Jones, W. K. (1968). Use and Regulation of the Radio Spectrum: Report on a Conference. *Washington University Law Review, 1968*(1), 71–115.
Legutko, C. (2008). Opinion: Changing the Regulatory Paradigm. *PolicyTracker*(June). Retrieved September 30, 2013, from www.policytracker.com
Levin, H. J. (1971). *The invisible RESOURCE*. Johns Hopkins Press.
Lie, E. (2004). *Radio Spectrum Management for a Converging World* Workshop on Radio Spectrum Management for a Converging World, Geneva.
London Economics. (2008). *Economic Impacts of Increased Flexibility and Liberalisation in European Spectrum Management: Report for A Group of European Communications Sector Companies*. http://londoneconomics.co.uk/

Louis, J. (2011). *International radio spectrum management beyond service harmonisation*. In Fourth International Conference on Emerging Trends in Engineering & Technology.
Louis, J., & Mallalieu, K. (2007). *Investigating the impact of convergence on the international spectrum regulatory framework*. In Proceedings of the Second International Conference on Systems and Networks Communications ICSNC '07, Washington D.C.
Lyall, F. (2011). *International communications: The international telecommunication union and the universal postal union*. Ashgate Publishing Ltd.
Man, P. D. (2016). *Exclusive use in an inclusive environment the meaning of the non-appropriation principle for space resource exploitation*. Springer International Publishing.
Marsden, R., Sexton, E., & Siong, A. (2010). *Fixed or flexible? A survey of 2.6 GHz spectrum awards*. DotEcon Discussion Papers.
Mazar, H. (2009). *An analysis of regulatory frameworks for wireless communications, societal concerns and risk: The case of radio frequency (RF) allocation and licensing*. Middlesex University. PhD Thesis.
O'Leary, T., Puigrefagut, E., & Sami, W. (2006). Overview of the Second Session (RRC-06) and the Main Features for Broadcasters. *EBU Technical Review* (October).
Radiocommunication Bureau. (2007). Report of the Director on the Activities of the Radiocommunication Sector on Resolution 951. In *World Radiocommunication Conference (WRC-07)*.
RCC. (2011). *Common Proposals by the RCC Administartions on WRC-12 Agenda Item 1.2* World Radiocommunication Conference (WRC-12), Geneva. Retrieved from https://www.itu.int
RCC. (2014). *Preliminary position of the RCC CA's on WRC-15 Agenda Items*. Retrieved from http://www.en.rcc.org.ru/
RCC. (2015). *Common Proposals by the RCC Administartions on WRC-12 Agenda Item 1.1* World Radiocommunication Conference (WRC-15), Geneva.
Roetter, M. F. (2009). The European Commission's Illogical 2.6GHz Decision. In *Responses to the Public consultation on the draft RSPG Work Programme 2010*.
RSPG. (2005). *RSPG05–102 on Wireless Access Platform for Electronic Communication Services (WAPECS) (A More Flexible Spectrum Management Approach)* (Radio Spectrum Policy Group Opinions, Issue. http://rspg.groups.eu.int
Ryan, P. S. (2005). The Future of the ITU and its Standard-Setting Functions in Spectrum Management. In S. Bolin (Ed.), *Standard Edge: Future Generation*. Sheridan Books.
Ryan, P. S. (2012). The ITU and the internet's titanic moment. *Stanford Technology Law Review, 2012*(8), 1–36.
Savage, J. (1989). *The politics of international telecommunications regulation*. Westview Press.
Selek, Y. K. (2008). *From 'Command and Control' Methods to 'Flexible' Spectrum Management: Comparing Turkey with other Markets*. University of Westminster. London.
Sims, M. (2007). WRC-07: The technological and market pressures for flexible spectrum access. *Communications and Strategies, 67*, 13–27.
Singh, R., & Raja, S. (2008). *Convergence in ICT services: Emerging regulatory responses to multiple play*. Retrieved from https://documents.worldbank.org
Standeford, D. (2013). African countries sign digital switchover agreement. *Policy Tracker*. Retrieved September 15, 2014, from https://www.policytracker.com
Struzak, R. (2007). *Spectrum management & regulatory issues*. In ITU Workshop on Market Mechanisms for Spectrum Management, Trieste.
Sutherland, E. (2006). *European Spectrum management: Successes, failures & lessons*. In ITU Workshop on Market Mechanisms for Spectrum Management, Geneva.
Sweden. (2014). *General Consideration Regarding Allocation and Use* Meeting of Fifth Meeting of Joint Task Group 4–5–6–7, Geneva.
United States Department of State. (2003). *United States Delegation Report: World Radiocommunication Conference 2003*. Retrieved from https://www.ntia.doc.gov/
USA. (2003). *World Radiocommunications Conference 2003: Proposals for the Work of the Conference*. Retrieved from https://www.itu.int

USA. (2015). *World Radiocommunications Conference 2015: Proposals for the Work of the Conference. Agenda Item 1.1.* Retrieved from https://www.itu.int

Watson, J. (2013). LTE Broadcast May Offer Mobile Operators a More Efficient Use of Spectrum. *PolicyTracker.com.* Retrieved 30/10/2013, from https://www.policytracker.com

Woolley, F. M. (1995). International frequency regulation and planning. *EBU Technical Review, Spring.*

Woolley, M. (1993). Working with WARC-92: Pragmatism and reciprocity needed. *Telecommunication Journal, 60*(1), 17–20.

Zacher, M. W. (1996). *Governing global networks: International regimes for transportation and communications.* Cambridge University Press.

Chapter 5
Technology Selection

> *The ITU is currently working on one of its most ambitious projects ever: systems standards for third generation mobile telecommunications...coined IMT-2000, it will make it possible to communicate anywhere anytime.*
>
> The Fifteenth Plenipotentiary Conference of the ITU (1998)

5.1 Introduction

Technology selection is a key element of spectrum management where the regulator may allow technology neutrality or predetermine particular technologies to be deployed by operators as in the case of the 'command and control' approach. With respect to personal mobile cellular communication, which is the focus of this chapter, the ITU-R has been involved in two aspects under the label IMT, namely defining the radio interfaces for IMT and the identification of spectrum of these IMT systems within the mobile service. This was mainly in response to the lack of interoperability between 2G standards and to achieve global roaming. In particular, the ITU-R work on IMT standards has resulted in six IMT-2000, two IMT-Advanced and three IMT-2020 terrestrial radio interfaces. In addition, over 17 GHz of frequencies have been identified for the use of IMT radio interfaces.

It is, however, unclear how the international activities of the ITU-R related to mobile technologies influence the decisions of national regulators when it comes to these technologies. Apart from national circumstances, such as market structure and auction design, the ITU-R work on IMT has an influence nationally on technology selection. It is, however, less clear when it comes to the standardization activities in the ITU-R, whose possible influence has largely overlooked. With this in mind, this chapter seeks to shed light on how IMT and national radio spectrum management policies interact with regard to technology selection. It starts with an overview of technology neutrality and discusses its application. The chapter then highlights the role of ITU-R in standardization with a focus on the case of cellular

M. A. El-Moghazi and J. Whalley, *The International Radio Regulations*,
https://doi.org/10.1007/978-3-030-88571-7_5

mobile telecommunication. The chapter then moves onto exploring IMT standardization and IMT spectrum identification. Finally, the chapter highlights the influence of IMT standardization on national spectrum policies with regard to technologies.

5.2 Technology Neutrality

Technology selection is one of the main elements of any spectrum management approach, where the selection of technologies can be neutral, restricted to standardized technologies or selective of specific technologies (ITU-R). Standardization refers to the level of specification of allocated services such as transmitter power and channelization, whereas technology neutrality is defined as the minimum applied constraints while ensuring that interference is appropriately addressed (Ali, 2009; Bohlin, 2012; Frullone, 2007; Kamecke & Korber, 2006). Another definition suggests that true technology neutrality implies defining conditions without any biased assumptions (Chaduc & Pogorel, 2008; Freyens, 2009).

There are pros and cons associated with limiting the choice of technology to standardized ones (Foster, 2008). The main advantage of standardization is that it allows for the large scale of production of equipment that reduces its cost. Moreover, the benefits of standardization include avoiding harmful interference and promoting interoperability between terminals and public networks. On the other hand, standardization may lead to locking operators and customers into an inferior standard and could delay the introduction of new equipment. An example of the failure in selecting the right technology is the enhanced radio messaging system (ERMES), which was an initiative to create a Europe-wide mobile messaging system that ended without any significant deployment of the technology (Whittaker, 2002). In fact, one of the main providers of ERMES, Ericsson, considers the standard to be a failure (Indepen & Aegis Systems, 2004; London Economics, 2008; Pogorel, 2007). It is argued that, in general, governments are not successful when it comes to selecting winning standards due to the absence of the industry's commercial motivation to identify efficient technologies (Cave, 2002).

With regard to personal mobile telecommunication, there have always been tensions between different technologies. For instance, in the 2G era, there was global competition between the GSM and CDMA standards, with the former winning in terms of the number of global subscribers (Bohlin, 2012; Cave et al., 2007; Pogorel, 2007). To be accurate, in some cases, there was no competition against GSM—in Europe, CEPT targeted from the very beginning the creation of a common European mobile system and created a new committee that was given the name GSM (Haug, 2002). Of course, this was not the case in the USA where the government encouraged neutrality among 2G operators during this period, the FCC continued its historical approach to licensing by letting the market place decide which technology to implement in any particular licensed band (Zelmer, 2002). During the 3G era, the most prominent example of competition was between WCDMA and CDMA2000 1xEV-DO/DV, with the former achieving a greater number of subscribers (Saugstrup &

Henten, 2006a). The race for 4G has been between the IEEE and 3GPP, which support WiMAX and the LTE-Advanced, respectively. This has ended with the later dominating the market as the single standard for 4G (Curwen, 2004; Saugstrup & Henten, 2006b).

The concept of technology neutrality has emerged with the development of personal mobile telecommunications. Initially, in Europe, for example, the European Commission (EC) enforced the exclusive use of the 900 and 1800 MHz bands for GSM (Saugstrup & Henten, 2006b). In the 3G era, El-Moghazi et al. (2015) suggest that the influence of the policymakers on technology development has decreased with the liberalization of the telecommunication markets. In particular, the calls for technology neutrality that started in the 1990s when the US government stated that there should be no assumed or required specific technologies, and that rules should be technology neutral (Curwen, 2004). Similarly, in Europe, technology neutrality has been one of the guiding principles of telecommunications regulation (Saugstrup & Henten, 2006b).

Although technology neutrality is a widely recognized and accepted concept (Ali, 2009), its practice of technology neutrality varies around the world. In Europe, GSM was mandatory for operators until 2009 (Lovells, 2014), when the use of spectrum bands reserved for 2G technologies was opened up to 3G technologies (Ali, 2009; Bohlin, 2012; Frullone, 2007; Kamecke & Korber, 2006; Sims, 2005). Japan took the position that the 2.5 GHz band would be reserved for non-IMT-2000 technologies (Cowhey et al., 2008), while in China, the government assigned each of the three operators a particular standard, namely TD-SCDMA, CDMA 2000 and WCDMA (Bohlin, 2012).

5.3 The Role of ITU-R in Standardization

The importance of having a harmonized global technical interconnection standard for wireless telecommunications emerged at the beginning of the twentieth century due to the incompatibility between maritime stations (Savage, 1989). Marconi forbade all stations using its equipment from communications with stations on ships of other manufactures. This was opposed by several countries who sought to safeguard their ships and prevent a monopoly emerging (Sims, 2006b). The Marconi Company refused to relay messages it received from competing operators (Newlands, 2009; Xia, 2011). Accordingly, an international conference was held in Berlin in 1903 to address the issue and to enable international interconnection (Savage, 1989). The final protocol of the conference established that wireless stations should operate in a way as not to interfere with other stations (Anker & Lemstra, 2011).

In addition, the International Long-distance Telephone Consultative Committee (CCIF), International Telegraph Consultative Committee (CCIT) and International Radio Consultative Committee (CCIR) were formed in 1925 and 1927, respectively, to establish international telecommunications standards for telephone, telegraph and radio (Zacher, 1996). However, when it comes to colour TV, it was not possible for

CCIR (now ITU-R) to agree on one standard in the 1960s. Instead, the American (NTSC), the French (SECAM) and the German/Dutch/British (PAL) standards were accepted (Zacher, 1996). It appears that, at this time, multiple standards were viewed by the ITU as the policy of last resort (Codding, 1991).

It has been argued that the ITU's role in setting international telecommunications standards will always be inferior to national and regional standards bodies for two key reasons (Besen & Farrell, 1991). Firstly, the operating procedures of these organizations are designed to be rapid and efficient, and the participants have enough influence in the market to enforce their agreements. Secondly, the processes of these organizations usually favour those with more market influence rather than, for instance, developing countries. Therefore, Besen and Farrell (1991) expected that the role of the ITU would be limited to 'rubber-stamping' the agreements of national and regional standardization organizations. But is such a claim valid when it comes to the role of the ITU-R in cellular mobile communications?

With the emergence of the commercial cellular mobile industry at the start of the 1980s, several different technological standards for 1G and 2G appeared (ITU, 2020). The ITU was motivated to intervene in the standardization process in response to the emergence of several national and regional standards bodies that were perceived as a threat to the role of the ITU in setting international telecommunications standards (Savage, 1989). More specifically, the lack of interoperability between 2G mobile standards motivated the ITU-R to become more involved in the standardization process of mobile systems (ITU-R, 2007c). It was also clear that the mobile market had become large and worldwide rather than national or regional in geographical scope (Besen & Farrell, 1991), which called for a global effort towards achieving a worldwide 3G standard.

The first step in the ITU standardization activities on mobile technologies was in the 1980s when Future Public Land Mobile Telecommunication System (FPLMTS) was discussed in the ITU-R (formerly CCIR). This occurred largely due to European countries seeking a successor to GSM (ITU, 1998). ITU-R Task Group 8/1 was formed in 1985 to define a framework for FPLMTS services (Funk, 1998). FPLMTS was renamed IMT-2000 by the Radiocommunication Assembly (RA) prior to the WARC-97 (U.S. Congress Office of Technology Assessment, 1993). WARC-92 also adopted Resolution 212, providing the general framework for IMT-2000 standards development and system implementation (Callendar, 1994). Since then, the ITU-R has been involved in different aspects of personal mobile communications systems or IMT (Blust, 2008). In the following two sections, we will focus on two key aspects, namely IMT standardization and IMT spectrum identification.

5.4 IMT Standardization

The main step for agreeing on the IMT-2000 radio interfaces was the definition of minimum requirements for technology evaluation. These were, even though the IMT-2000 sought to provide a data rate of up to 2 Mbit/s (Engelman, 1998):

- Indoor office—supporting a bit rate of at least 2048 kbit/s at low speeds of movement;
- Outdoor to indoor and pedestrian—supporting a bit rate of at least 384 kbit/s at medium speeds;
- Vehicular—supporting a bit rate of at least 144 kbit/s at high speeds (ITU-R, 2003a).

The ITU-R established a procedure for submitting and evaluating the IMT-2000 radio interfaces proposed by national and regional standardization bodies (ITU, 1997). The ITU-R then invited in 1997 applicants for IMT-2000 radio transmission technologies to apply, with ten terrestrial proposals being submitted in the following year (Leite et al., 1997). Among these proposals, five radio interfaces were approved to be part of the IMT-2000 standards: WCDMA (also known as UMTS), CDMA 2000, TD-SCDMA, EDGE and DECT (ITU-R, 2010). The RA of 2000 approved the detailed specifications of the radio interfaces for IMT-2000 in ITU-R Recommendation M.1457 (ITU-R, 2003a).

It is worth highlighting that the ITU vision during the early stages of IMT standardization was to select a single global standard for mobile communications mainly to achieve economies of scale. The main goal was summarized as *'communicate anywhere–anytime offering a seamless operation of mobile terminals worldwide'* (ITU, 1998). Initially, there were views that this one global standard should be WCDMA (Cowhey et al., 2008). However, it was not possible to obtain consensus on this, and the process of choosing technologies for IMT-2000 ended up with several technologies instead of one. This partially reflected the impact of intellectual property issues, the support of the USA for technology neutrality (Cowhey et al., 2008) and the inability to agree on a single domestic standard in Japan (Funk, 1998). As a consequence, the ITU eventually focused on harmonizing the different 3G standards instead of having a single standard (Georgieva, 2008).

In 2006, the IEEE submitted a proposal to include the WiMAX standard IP-OFDMA into the IMT-2000 family of standards (ITU-R, 2009). This was opposed by China, Germany and several industry bodies (WP 5D Chairman, 2007a). These industry bodies, representing competing technologies, felt that the compliance of WiMAX with the minimum performance capabilities of IMT-2000 needed further work. Therefore, a special meeting was held in August 2007 to address the concerns related to the inclusion of WiMAX (WP 5D Chairman, 2007b). The meeting ended up with Germany and China still having some reservations but with the majority of the attendants supporting the inclusion. The ITU-R RA-07 agreed in October 2007 to officially include WiMAX into the IMT-2000 family (ITU-R, 1997). While the IMT-2000 standardization process ended up with six different technology standards, only three of them were prominent when it came to market deployment: TD-SCDMA, CDMA 2000 and WCDMA (Blust, 2012).

As the 4G era loomed, the ITU–R announced its invitation for the submission of proposals for candidate radio interface technologies for the terrestrial components of the successive systems to IMT-2000, which were named 'IMT-Advanced'

(WIMAXForum, 2007). The key feature of IMT-Advanced was set to be the provision of enhanced peak data rates of up to 100 Mbit/s for high mobility and 1 Gbit/s for low mobility (Sims, 2007b). There were six different proposals containing two main technologies: IEEE (IEEE 802.16 m) and 3rd Generation Partnership Project (3GPP) (LTE Release 10) (ITU, 2008). Later, these six proposals were consolidated into two IMT-Advanced technologies: LTE-Advanced and WirelessMAN-Advanced (ITU-R, 2007a). These two technology standards were submitted to the RA-12 and were agreed by ITU-R Member States (WP 5D Chairman, 2009). The ITU-R issued recommendation ITU-R M.2012 that accommodates the specifications of these two standards (ITU-R, 2012b). The LTE-Advanced standard was identified as LTE Release 10 and beyond, while the WirelessMAN-Advanced standard was identified as IEEE 802.16 m (ITU, 2010a). Ultimately, WiMAX Advanced rather than LTE-Advanced dominated the market (Blust, 2012). It is worth mentioning that there is a satellite component of IMT which encompasses both IMT-2000 and IMT-Advanced as identified in Recommendation ITU-R M.1850 and Recommendation ITU-R M.2047, respectively (ITU-R). However, the focus of this chapter is on the terrestrial components of IMT.

Regarding 5G standards, the ITU-R developed a framework and set of overarching objectives for the future development of what is called ‘IMT for 2020 and beyond’ in 2012 (ITU, 2012). IMT-2020 accommodates three usage scenarios of IMT-2020, which are shown in Fig. 5.1. These are as follows:

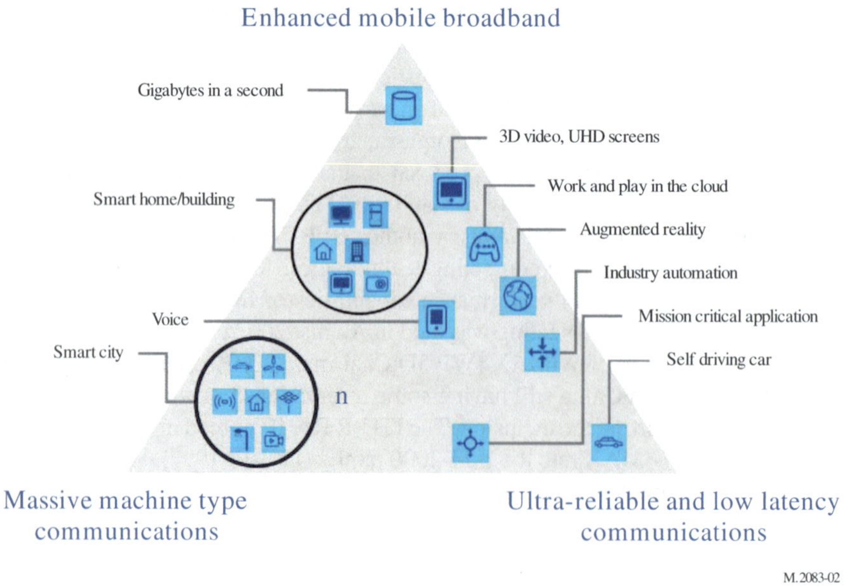

Enhanced mobile broadband

Gigabytes in a second

3D video, UHD screens

Smart home/building

Work and play in the cloud

Augmented reality

Industry automation

Voice

Mission critical application

Smart city

Self driving car

Massive machine type communications

Ultra-reliable and low latency communications

M.2083-02

Fig. 5.1 IMT usage scenarios. *Source* ITU-R (2015c)

- enhanced mobile broadband; human-centric use cases for access to multimedia content, services and data;
- massive machine type communications; a large number of connected devices typically transmitting a relatively low volume of non-delay-sensitive data;
- ultra-reliable and low latency communications; stringent requirements for capabilities such as throughput, latency and availability.

One key step was for ITU-R, in March 2016, to issue a Circular Letter inviting submissions of proposals for candidate radio interface technologies for the terrestrial components of the radio interface(s) for IMT-2020. The same letter also invited participation in the subsequent evaluation of the proposals, with a scheduled start date of October 2017 (ITU-R, 2016a). The submission and evaluation process of the IMT-2020 was organized via several ITU-R reports. Report ITU-R M.2411, for example, provided the evaluation criteria and submission templates for IMT-2020, while Report ITU-R M.2412 outlined the guidelines for evaluating radio interface technologies. In February 2017, the ITU agreed on key 5G performance requirements for IMT-2020 (ITU, 2017b). The IMT-2020 key capabilities are different in many aspects to IMT-Advanced, most notably IMT-2020 which has a peak data rate of 20 Gbit/s compared to 1 Gbit/s and user-experienced data rate of 100 Mbit/s compared to 10 Mbit/s in IMT-Advanced. With respect to devices connection density and latency, IMT-2020 may have up to 1 million devices and 1 ms. The improved performance of IMT-2020 over IMT-Advanced is evident in Fig. 5.2.

By the end of 2019, seven candidate IMT-2020 proposals had been submitted and received evaluation reports from the independent evaluation groups (ITU-R, 2019a). Two of these were the 3GPP submissions for IMT-2020, which were independent of one another. NR alone was submitted as a Radio Interface Technology (RIT) proposal for IMT-2020, while NR and E-UTRA/LTE were jointly submitted as two component RITs of a set of radio interface. 5G New Radio (NR) is the 3GPP name for the 5G radio interface (5GAmericas, 2020). China and Korea submitted another two proposals, which were identical to 3GPP RIT submissions, with the remaining three proposals being DECT SRIT, Nufront RIT and TSDSI RIT. 13 independent groups evaluated these proposals (ITU-R, 2020a). The work on IMT-2020 was expected to be finalized in 2020 with the publishing of the specifications for the new IMT-2020 radio interfaces (ITU, 2017a).

The evaluation of candidate IMT-2020 radio interfaces recently ended in February 2021 with the approval of two 3GPP technologies (5G-SRIT and 5G-RIT) and TSDSI technology (5Gi), with the latter being largely based on 3GPP-NR and an additional component (LMLC) to provide low-cost 5G in rural areas (ITU-R, 2021b). Two submissions were evaluated as being inconclusive, namely DECT SRIT and Nufront RIT. However, it was decided that these two submissions required additional evaluations to conclude their assessments and thus will be granted an exceptional extension to provide additional material (ITU-R, 2020b). The work towards future technology trends for 'IMT towards 2030 and beyond' started in 2021 to anticipate new use cases for IMT (ITU-R, 2021a). Table 5.1 summarizes a timeline of the ITU-R activities regarding IMT standardization.

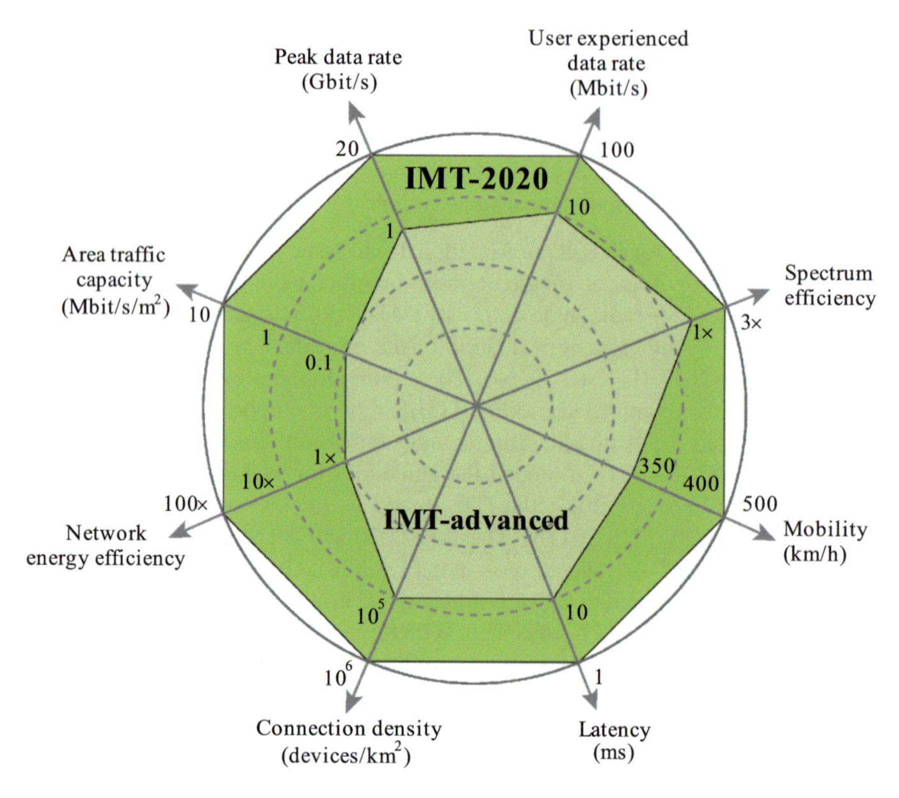

Fig. 5.2 IMT-2020 versus IMT-advanced. *Source* ITU-R (2015d)

Table 5.1 Timeline of IMT standardization

Year	Standardization activity
1985	The ITU-R Task Group 8/1 was formed to define a framework for FPLMTS services
1997	The name FPLMTS was renamed IMT-2000
1997	The ITU-R invited applicants for IMT-2000
1998	Ten terrestrial proposals were submitted for IMT-2000
1999	Five radio interfaces were approved to be part of the IMT-2000 standards
2006	The IEEE submitted a proposal to include WiMAX into the IMT-2000
2007	RA-07 agreed to include WiMAX to the IMT-2000 family
2008	The ITU–R announced its invitation for the submission of applicants for IMT-advanced
2009	Six different proposals containing two main technologies (IEEE 802.16 m and LTE Release 10) were submitted
2012	The RA-12 approved the two technologies to be part of the IMT-Advanced
2016	Start of IMT-2020 circular letter preparation
2020	Finalization of IMT-2020 technologies specifications

5.5 IMT Spectrum Identification

The IMT standardization process has been associated with identifying the radio spectrum bands to be used by standards. The idea of identification is related to the problem of having multiple incompatible 2G standards with different frequency bands and frequency arrangements (Rancy, 2017). The first step in the IMT spectrum identification process was when the European countries called for spectrum allocation for FPLMTS during the mobile WARC of 1987. This issue was, however, postponed until WARC-92. During WARC-92, countries such as the USA opposed allocating new frequencies for FPLMTS (Osseiran, 2013). Though eventually the conference identified the 1885–2025 MHz and 2110–2200 MHz bands for countries wishing to implement FPLMTS (ITU-R, 2016b). This amount of identified spectrum was related to the estimated spectrum requirements at that time, which were around 230 MHz (ITU-R, 2003b). Furthermore, WARC-92 upgraded the service status in the 1700–2690 MHz band to primary for the mobile service within which FPLMTS can operate (ITU, 1992). As the main idea of identification is that it does not preclude the use of the additional bands for other types of mobile applications or by other services to which these bands are allocated, this helped persuade countries to agree to the concept (ITU, 2000c).

As the identification concept was not perceived as being obligatory at that time, the USA deployed PCS, a 2G mobile technology, in frequencies identified for IMT with the consequence that operating IMT systems in these frequencies was not possible (ITU, 2000a). The USA assigned the 1850–1950 MHz frequency band for Personal Communications Services (PCS-1900) utilizing an adopted version of GSM (Huurdeman, 2003). That may explain why the USA proposed no mobile allocations and designations for FPLMTS, given that their internal regulations provided flexibility and that the designation of spectrum for FPLMTS was premature (U.S. Congress—Office of Technology Assessment, 1991). Furthermore, the USA opposed FPLMTS because it was believed that the current allocation was sufficient to provide the service and that their focus at that time was on satellite-based services. In addition, the definition of FPLMTS was not considered to be accurate nor clear with the consequence that flexibility when using the spectrum may decrease (U.S. Congress—Office of Technology Assessment, 1993). An interesting related incident occurred when the EU objected to the USA implementing PCS in frequencies already identified for IMT, considering this to be a violation of the ITU RR. The USA, however, felt that IMT identification did not have any regulatory status (Manner, 2003).

It should be noted that the name 'IMT-2000' was chosen to reflect that the service would use the 2 GHz spectrum band and be available in the 2000s (Jho, 2007). El-Moghazi et al. (2014) argued that there was an indirect link between spectrum and IMT standards. In other words, the IMT-2000 sent signals to the market that 3G would be deployed in the 2 GHz band. They further argue that this could be one of the reasons why the 3G licenses that were awarded at the early 2000s received large bids as a result of the fear of operators that they will not be able to obtain the required spectrum to provide 3G services. In any case, it seems that IMT spectrum identification, along with the approval of IMT standards, stimulated licensing in the 2 GHz band with several commercial operations commencing (ITU, 2000b).

The ITU work on IMT identification also enabled the refarming of the band previously utilized for 3G to IMT-2000. That was enabled by the WRC-2000 decision to identify the 806–960 MHz band that was already allocated to mobile service to IMT-2000 (ITU-R, 2001a). Such a decision intended to enable the operation of 1G and 2G mobile services in bands identified to IMT-2000 (ITU-R, 2001b). Furthermore, the ITU-R also contributed to achieving the long delayed harmonization between the ITU-R three regions in the UHF band when WRC-12 allocated the 694–790 MHz band to the mobile except aeronautical mobile service in Region 1 (ITU-R, 2012c). This resolved the historical incompatible deployment of CDMA and GSM networks in the UHF band globally (ITU-D, 2012). El-Moghazi et al. (2014) clarified that the ITU-R decision to have the IMT label include IMT-2000 and IMT-Advanced allowed both systems to access all of the spectrum identified for IMT. In other words, both 3G and 4G systems can access the same spectrum.

With respect to 5G frequencies, WRC-19 addressed the potential IMT identification in the millimetre bands under Agenda Item 1.13, which considered the identification of frequency bands for the future development of IMT in accordance with Resolution 238 of WRC-15 (ITU-R, 2016c). Such a resolution is related to the frequency for IMT identification and included the possible additional allocations to mobile services on a primary basis in parts of the frequency range between 24.25 and 86 GHz. The focus of the conducted studies has been on the frequency bands (e.g. 24.25–27.5 GHz, 47.2–50.2 GHz and 81–86 GHz bands) that have allocations to mobile services on a primary basis, as well as those bands (e.g. 40.5–42.5 GHz and 47–47.2 GHz bands) which may require additional allocations of mobile services on a primary basis. Eventually, WRC-19 identified three bands on a global basis for IMT (24.25–27.5 GHz, 37–43.5 GHz and 66–71 GHz) and other bands in some countries via footnotes (45.5–47 GHz and 47.2–48.2). Table 5.2 provides a timeline of the IMT spectrum identification by different WRCs.

It is useful to highlight the importance of having IMT identification rather than operating only within the mobile allocation. In terms of the advantages of identification, those developing countries (e.g. Africa) that rely heavily on the RR prefer

Table 5.2 Timeline of IMT spectrum identification

Conference	IMT identified bands
WARC-1992	1885–2025 MHz and 2110–2200 MHz
WRC-2000	806–960 MHz, 1710–1850 MHz, 2500–2690 MHz
WRC-2007	450–470 MHz, 698–806/862 MHz, 2.3–2.4 GHz and 3.4–3.6 GHz
WRC-2012	694–790 MHz
WRC-2015	470–694 MHz, 1427–1518 MHz, 3300–3400 MHz, 3400–3600 MHz, 3600–3700 MHz and 4800–4990 MHz
WRC-2019	24.25–27.5 GHz, 37–43.5 GHz, 45.5–47 GHz, 47.2–48.2 and 66–71 GHz

Source ITU-R (2001a, 2008, 2012a, 2015a); U.S. Congress Office of Technology Assessment (1993)

not to deploy IMT outside of the identified bands (SADC-EACO, 2019). Another reason for seeking the IMT identification is that in these countries the identified band for IMT is already occupied by legacy systems, and they draw extensively on the sharing and compatibility studies conducted by the ITU-R in order to have certainty when utilizing the band for IMT. This was apparent in the case of the C-band, where African countries are seeking to identify the band in WRC-23. From the industry viewpoint, sometimes there is a need for the RR to align with what is happening in the industry, especially where deployments have already occurred (GSMA, 2019).

Another reason for having bands identified for IMT is to make sure the existing service allocations are protected. This may explain why CEPT countries are keener to have identification in the 24 GHz band rather than in the C-band. In the case of the C-band, the regulator could establish exclusion areas around FSS earth stations, but the EESS deployed directly below the 24 GHz is based on receiving (listening) to large areas. For example, the European EESS system, which is called 'ESA', measures the temperature globally and thus is wider than just Europe. Therefore, there is a need for European countries to have their out of band emissions (OOBE) values from IMT towards EESS systems below 24 GHz incorporated in the RR and to convince other non-European countries to adopt such OOBE values. In other words, there is a need for the mobile operation in the 26 GHz to be aligned with the RR in all countries deploying IMT (Ercole, 2018). This is because the EESS is already protected by the RR and has several filings within the ITU-R frequencies database (MIFR). Thus, IMT identification in the band 26 GHz would imply protection for EESS in the 24 GHz band (Standeford, 2018).

On the other hand, operating without IMT identification would entail a higher risk of interference. In the C-band, there is a risk of potential satellite interference from neighbouring countries and military services (Marti, 2019). In particular, not identifying the 3.6–3.8 GHz band for IMT at WRC-15 resulted in no protection from existing or future satellite earth stations. While this may not be an issue if all countries harmonized IMT identification, this may not be the case for those European countries close to other countries having a different use of the C-band (e.g. Finland and Russia).

5.6 Influence of IMT on National Technology Policies

In general, there are different perspectives on the role of governments in standards setting. While the USA prefers to let the market handle the standardization process, China and the EU usually intervene (Biddle et al., 2012). This is also the case for the impact of the IMT standardization process on the decisions that regulators take when picking a technology. For some regulators, mandating that selection of technologies to be from the IMT standards family is an application of technology neutrality as no specific IMT standard is chosen. Other regulators, however, believe that neutrality means not imposing any particular technology including the IMT standards on operators. In all cases, the ITU does not have the authority to enforce any particular standard on any country. However, the ITU can still influence a regulator's decision.

In Europe, where the GSM standard was mandatory for operators (Cowhey et al., 2008), there was a discussion over whether those bands identified for IMT-2000 could be used for technologies other than IMT-2000. This was mainly due to pressure from the European Commission (EC) to apply technology neutrality. Furthermore, the Electronic Communications Committee (ECC), which is one of the CEPT bodies, clarified that designating a band for a particular use does not preclude adopting other technologies but rather implies that *'countries are obliged to make spectrum available for a harmonised application if market demand is great enough to exclude other services and applications'* (Sims, 2005). Eventually, the EC could not reach a consensus among its (then) Member States on allowing non-IMT-2000, including WiMAX, to use IMT-2000 identified spectrum such as the 2.5 GHz (Sims, 2006a). On the other hand, Japan took the position that the 2.5 GHz band will be reserved for non-IMT-2000 technologies (Sims, 2006b). Some countries were more specific regarding the selection of technology from the IMT standards family. In Hong Kong, the award of a standard specific CDMA2000 license was announced (Sims, 2007a), while in China, the government has assigned each of the three operators a different standard (Newlands, 2009).

Regarding the influence of IMT standardization on technology selection, El-Moghazi et al. (2019) argue that being part of the IMT family of standards provides support to the applicant technologies to achieve economies of scales and roaming. However, there is another view that being an IMT technology is a secondary priority to other factors that have a greater influence on the regulator's technological decision such as, for example, regional harmonization. El-Moghazi et al. (2019) also highlight the view that IMT standardization could encourage more neutrality when selecting a technology, and the IMT family of standards is considered to be in conformity with technology neutrality as the standardization process is open and transparent. The second view is that IMT standardization may discourage regulators from being neutral and thus may have a negative influence on operators and technology developers as it restricts the technologies to be adopted to IMT systems.

The IMT standardization process has also shown that the telecommunications industry and ITU-R have different perceptions regarding the categorization of technologies as being 2G, 3G or 4G. According to the ITU-R, IMT-2000 is equivalent to 3G, but this is not necessarily the case for industry. For instance, although EDGE is part of the IMT-2000 family of radio interfaces, it is not considered by industry as a 3G technology (ITU-D, 2009). Moreover, the first releases of CDMA2000 were considered by some as not being a real 3G technologies although it is part of the IMT-2000 family of standards (Saugstrup & Henten, 2006a).

The emergence of the IMT-Advanced family of standards caused a similar debate regarding the categorization of 3G and 4G technologies. The issue is that IMT-2000 standards encompass their enhancements and future developments (ITU-R, 2007b) and some of the IMT-2000 radio interfaces have evolved to achieve capabilities that are similar to the IMT-Advanced (ITU, 2010c). For instance, LTE Release 10 is part of the IMT-2000 radio interfaces, and it supports up to 3 Gbit/s in the downlink which is way beyond the initial capabilities of IMT-2000 (ITU-R, 2013b). Moreover, LTE Release 10 and 11 are also part of the IMT-Advanced radio interfaces (ITU-R, 2013a). It is worth mentioning that the IMT-Advanced standards have been called 'true 4G technologies' by the ITU (ITU, 2009).

The influences on defining what is commercial vary from country to another. For instance, in Japan, LTE is understood to be 3G. In contrast, T-Mobile in the US branded HSPA+ as 4G technology while Verizon considered LTE to be 4G (Goldman, 2010; Ziegler, 2011a). AT&T considered both LTE and HSPA+ to be 4G technologies (Ziegler, 2011b). It is worth noting that the ITU clarified in 2010 that while IMT-Advanced is considered as 4G, the term '4G' per se may also be applied to the forerunners of these technologies—that is, LTE and WiMAX—and to other evolved 3G technologies (ITU, 2010b).

With respect to the regulator's perspectives, in the UAE, there was a dispute between Etisalat and Du over who was the first to launch 4G services in the country. Du launched an advertisement campaign claiming it was the first to launch 4G services in the UAE by deploying HSPA+ technologies that provided speeds of up to 42 MB/sec. This technology is considered by the ITU to be a 4G service. Etisalat objected, arguing that HSPA+ is 3.75G rather than 4G according to TRA standards (emirates247, 2011b). Eventually, TRA instructed Du to suspend its services and to modify its declaration to run the 4G services. It was also worth mentioning that TRA found that the ITU did not issue a decision on the HSPA+ technology (emirates247, 2011a).

In Egypt, the first two mobile licences granted by NTRA were specific to 2G technologies, while the third mobile licence granted was technology neutral. This motivated the initial two mobile operators to acquire 3G licences (Aryani et al., 2009). Following the auction of the third license in 2006, there was conflict between NTRA and Mobinil. NTRA notified Mobinil to stop it introducing EDGE until it had acquired a 3G license. While NTRA considered EDGE as a technology capable of providing 3G services, Mobinil considered EDGE to be an improvement of GPRS networks and should not be classified as 3G. Moreover, GSMA disagreed with NTRA, arguing that EDGE is an improvement of GPRS and should not require a new license (Dardeery, 2006). The stance of NTRA was that the air interface of EDGE is different to GPRS and that, according to the ITU, EDGE was approved in 2000 as a 3G standard (NTRA of Egypt, 2006b). Eventually, Mobinil postponed the promotion of EDGE services until the disputes with NTRA were settled (NTRA of Egypt, 2006a). After almost a year, NTRA and Mobinil reached an agreement where Mobinil was granted a 3G license that allowed them to use UMTS, EDGE, HSUPA and HSDPA technologies and to acquire 10 MHz of broadcasting spectrum for its 3G services (ITU, 2007).

El-Moghazi et al. (2019) examined the influence of IMT standardization on generations of mobile technology definitions and showed that while one view is that there is mutual influence between the IMT definitions and definitions of 3G and 4G, the alternative view is that there is no influence because the ITU-R has taken several actions to diminish any influence of IMT standards on the technology generations definitions used in the market. They also suggested that IMT standardization has largely no influence on the discrimination between technology generations nationally. This is because regulators differentiate between technology generations due to reasons related to national telecommunication market circumstances rather than because of reasons related to IMT. Moreover, the ITU-R has also sought to reduce the variation between IMT-2000 and IMT-Advanced and not to be involved in the discussions related to the differences between 3 and 4G. However, regulators could use the IMT-2000 and IMT-Advanced labels to justify their discrimination between technologies.

As we have highlighted the importance of IMT identification in the previous sections, it is worth examining the influence of IMT spectrum identification on national spectrum policy with respect to technology selection. There are different perspectives on IMT identification and whether it is mandatory to have such identification to deploy IMT systems. This is important because before WRC-19 several countries and regions decided on which spectrum bands to be utilized by 5G technologies without waiting for the conference. This included bands that were rejected for IMT identification by WRC-15 such as the C-band (3.6–3.8 GHz), which has been adopted by many Region 1 countries as the pioneer band for 5G, and the middle of the L-band (1452–1492 MHz), which is not identified for IMT in CEPT countries. This also extends to bands that are still under study by WRC-19 where disagreements regarding deployment conditions exist (e.g. 24.25–27.5 GHz). Similarly, the 27.5–29.5 GHz band, which was omitted from the list of bands to be considered by WRC-19, has been decided to be a 5G band in several countries including the USA and South Korea.

To illustrate the different perspectives on IMT spectrum identification, one particular ITU-R recommendation related to the IMT frequency arrangement contains important provisions for IMT identification. It stated *'administrations may deploy IMT systems in bands allocated to the mobile service other than those identified in the RR'* (ITU-R, 2015b). Such a provision indicates that IMT identification is not a condition for deploying IMT system but rather that it entails that mobile service allocation is necessary. Russia submitted a contribution to WP 5D that refuted such an understanding and proposed to omit such a statement from the recommendation (Russian Federation, 2018). The contribution suggested that following WRC-15, which discussed additional spectrum for IMT systems in the range below 6 GHz, the situation in the RR has changed in terms of the format of the footnotes that identified frequency bands for IMT—that is, IMT became more specific in the footnotes. As stated in the contribution *'WRC-15 has allotted IMT into a separate type of communication systems, requiring additional regulation regardless of pre-existing mobile service allocation in which the IMT operates'* (Russian Federation, 2018). The contribution also highlighted that following WRC-15, the ITU-R BR has introduced a new symbol for 'Nature of Service' specifically for IMT in order to, firstly, distinguish the frequency assignments to IMT stations from other stations in the mobile service and, secondly, enable the examination of the conditions associated with IMT for the purpose of notifying assignment in the MIFR. Therefore, with this in mind, before

WRC-15, having IMT operating in bands allocated to mobile services does not imply additional obligations.

There was a quite tense discussion on such a provision in the 2019 Radio Assembly that ended up with modifying the provision to indicate that the operation of IMT in bands allocated to mobile services but not identified to IMT should comply with the RR and the ITU-R recommendations (ITU-R, 2019b). Significantly, the text that was included was drafted as a compromise: *Also, administrations may deploy IMT systems in bands allocated to the mobile service other than those identified in the RR, and administrations may deploy IMT systems only in some or parts of the bands identified for IMT in the RR. However, it is emphasized that the use of IMT in any band allocated to the mobile service on a primary basis but not identified for IMT should also comply with the objectives of the relevant technical and regulatory provisions of the RR, as well as with the latest version of applicable ITU-R Recommendation(s)* (ITU-R, 2015d).

Another example of different perspectives on IMT identification is demonstrated by the L-band, involving countries deploying IMT without having their names included in the footnote identifying the band for L-Band. In particular, the Russian Federation was concerned that the use of IMT stations should be limited to only those countries mentioned in the footnotes, if not, those countries which do not have international recognition for their stations may cause significant interference to existing services in other countries (Russian Federation, 2018).

The various stances regarding whether IMT identification is required for mobile use deployment are also related to whether IMT identification implies exclusive access to IMT or simply sharing. As previously mentioned, IMT identification does not preclude other uses of the spectrum and should not contradict existing service allocations. In other words, the identification indicates specific use under the mobile service allocation for the IMT systems as defined by the ITU-R recommendation. This means that it does not necessarily imply only utilizing the spectrum for IMT on an exclusive basis.

El-Moghazi and Whalley (2019) argue that IMT identification is necessary for cellular mobile deployment in some countries while for other countries (e.g. USA), mobile service allocation is sufficient. In both cases, the countries operate within the ITU-R framework but with a different interpretation, and having IMT identification is a priority even for those who operate without such an identification. This is because operating without IMT identification is associated with uncertainty in terms of protection against interference and the legal status of services. There are, however, still benefits of operating outside of IMT identification such as the avoidance of strict sharing conditions that add to the cost of deployment. However, operating without IMT identification comes with a price that may, in some cases, be too expensive. For example, countries operating 5G in the C-band have no protection against interference from FSS in neighbouring countries. Furthermore, there were concerns before WRC-19 that the deployment conditions of 5G in the 26 GHz band could have been challenged at the conference. This could have added to the uncertainty in terms of protection against interference and legal status and may have caused additional deployment cost.

To understand another perspective on the IMT identification, it is useful to highlight the positions of broadcasters with respect to Footnote 5.295 of the RR that proposes IMT identification in the 470–608 MHz band in the USA. Following the incentive auction in 2017, the mobile operators obtained access to the 614–698 MHz band, while broadcasters retained access to the 470–608 MHz band. The broadcasters demanded that the FCC removes the USA from the footnote as it did not reflect the reality emerging subsequent to the auction (Youell, 2018).

El-Moghazi et al. (2019) examined the influence of IMT spectrum identification on technology selection and neutrality, finding that such an identification had a positive impact on a regulator's tendency to select a technology from the IMT family. However, the extent of such influence is dependent on the country. Identification makes spectrum more valuable, and having a particular band identified for IMT supports the regulator when they want to refarm spectrum to another use. Furthermore, as El-Moghazi et al (2019) demonstrate, IMT spectrum identification contains elements of flexibility and restriction that impact on technology neutrality. Firstly, restrictive elements, including having spectrum identified to IMT, may imply using the spectrum for IMT. This, in turn, may discourage regulators from applying technology neutrality and to limit the use of the band to only IMT. Secondly, the flexible elements included are such that identification does not preclude other mobile use and that mobile development has been progressed in bands other than those identified for IMT.

5.7 Summary

In this chapter, we have addressed technology selection. This is one of the main elements of any spectrum management approach, where the selection of technologies can be neutral, restricted to standardized technologies or selective of specific technologies.

With regard to personal mobile telecommunication, in the 2G era, there was global competition between GSM and CDMA, with the former winning in terms of the number of subscribers. During the 3G era, the most prominent example of competition was between WCDMA and CDMA2000 1xEV-DO/DV, with the former achieving a greater number of subscribers. The race for 4G between the IEEE and 3GPP, which support WiMAX and the LTE advanced respectively, has ended with the later dominating the market as the single standard for 4G.

Meanwhile, the lack of interoperability between 2G mobile standards motivated the ITU-R to become more involved in the standardization process of mobile systems, with the first step being when IMT-2000 (formerly FPLMTS) was discussed in the 1980s. Since then, the ITU-R has been involved in two aspects under the label IMT: defining the radio interfaces for IMT and identifying spectrum for these IMT systems within the mobile service. In particular, the ITU-R established a procedure for submitting and evaluating the IMT-2000 radio interfaces proposed by national and regional standardization bodies, and these procedures have resulted in having a number of terrestrial radio interfaces across three families of standards.

The chapter shows that being part of the IMT family of standards provides support to the applicant technologies to achieve economies of scales and roaming. There is, however, another view that being an IMT technology is of secondary importance to other factors that have a greater influence on regulator decision on technology such as regional harmonization. Regarding the influence of IMT on technology neutrality, there is a view that IMT standardization could encourage more neutrality when selecting a technology. The IMT family of standards is considered to be in conformity with technology neutrality due to the process being open and transparent. The second view is that the IMT standardization may discourage regulators from being neutral and thus may have a negative influence on operators and technology developers as it restricts technologies adoption to IMT systems.

The IMT standardization process has also shown that the industry and ITU-R have different stances on the categorization of technologies as 2G, 3G or 4G. While one view is that there is mutual influence between the IMT definitions and the definitions of 3G and 4G, the alternative view is that there is no influence because the ITU-R has sought to diminish any influence of IMT standards on the technology generations definitions used in the market.

Finally, this chapter also revealed that IMT identification is necessary for cellular mobile deployment in some countries while for other countries (e.g. USA), mobile service allocation is sufficient. In both cases, the countries operate within the ITU-R framework but with a different interpretation, and having IMT identification is a priority even for those who operate without such an identification. Regarding the influence of IMT spectrum identification on technology selection and technology neutrality and showed that such identification has a positive impact on a regulator's tendency to select a technology from the IMT family. However, the extent of such influence is dependent on the country.

References

5GAmericas. (2020). *The 5G evolution: 3GPP releases 16-17*. www.5gamericas.org

Ali, R. (2009). Technological neutrality. *Lex Electronica, 14*(2), 1–15.

Anker, P., & Lemstra, W. (2011). Governance of radio spectrum: License exempt devices. In W. Lemstra, V. Hayes, & J. Groenewegen (Eds.), *The innovation journey of Wi-Fi: The road to global success*. Cambridge University Press.

Aryani, L., Cankorel, T., Chadbourne, & LLP, P. (2009). Regulating WiMAX to life in the Middle East and North Africa. *Telecomfinance* (November). www.telecomfinance.com

Besen, S. M., & Farrell, J. (1991). The role of the ITU in standardization: Pre-eminence, impotence or rubber stamp? *Telecommunications Policy, 15*(4), 311–321.

Biddle, B., Curci, F. X., Haslach, T., Marchant, G. E., Askland, A., & Gaudet, L. (2012). The expanding role and importance of standards in the information and communications technology industry. *Jurimetrics, 52*, 177–208.

Blust, S. (2012). Global deployments of technologies utilizing IMT specifications and standards. WP5A-WP5B-WP5C Workshop on the Preparations for WRC-15, Geneva.

Blust, S. M. (2008). Development of IMT-advanced: The SMaRT approach. *ITU News*(10).

Bohlin, E. (2012). Technology Neutrality—Why Europe got on board. In *International Roundtable Discussion on a Technology Neutral Spectrum Regime in the Indonesian Cellular Business*, Jakarta.

Callendar, M. H. (1994). Future public land mobile telecommunication systems (FPLMTS). In *The Third Annual International Conference on Universal Personal Communications*, San Diego.

Cave, M. (2002). *Review of radio spectrum management, an independent review for department of trade and industry and HM treasury*. www.ofcom.org.uk

Cave, M., Doyle, C., & Webb, W. (2007). *Essentials of modern spectrum management*. Cambridge University Press.

Chaduc, J., & Pogorel, G. (2008). *The radio spectrum. Managing a strategic resource*. ISTE Ltd.

Codding, G. A. (1991). Evolution of the ITU. *Telecommunications Policy, 15*(4), 271–285.

Cowhey, P. F., Aronson, J. D., & Richards, J. E. (2008). The peculiar evolution of 3G wireless networks. In W. J. D. A. E. J. Wilson (Ed.), *Governing global electronic networks: International perspectives on policy and power*. MIT Press.

Curwen, P. (2004). Has GSM won the technology war? *Info, 6*(1), 61–62.

Dardeery, H. (2006). Crisis escalates between NTRA and MobiNil over EDGE technology. *Daily News Egypt*. Retrieved on 30 November, 2014, from https://dailynewsegypt.com

El-Moghazi, M., & Whalley, J. (2019). *IMT spectrum identification: Obstacle for 5G deployments*. TPRC, Washington, D.C.

el-Moghazi, M., Whalley, J., & Irvine, J. (2014). IMT standardisation and spectrum identification: Regulatory and technology implications. In *ITU Kaleidoscope Academic Conference*, Saint Petersburg.

El-Moghazi, M., Whalley, J., & Irvine, J. (2015). Influence of IMT standardisation process on WiMAX decline. In *Second Regional ITS Conference*, New Delhi.

El-Moghazi, M., Whalley, J., & Irvine, J. (2019). Technology neutrality: Exploring the interaction between international mobile telecommunication and national spectrum management policies. *Telecommunications Policy, 43*(6), 531–548.

emirates247. (2011a). Du told to suspend 4G. *emirates247*. Retrieved on 30 November, 2014, from https://www.emirates247.com

emirates247. (2011b). Who first launched 4G in UAE—Du or Etisalat? *emirates247*. Retrieved on 30 November, 2014, from https://www.emirates247.com

Engelman, R. B. (1998). Keynote Address. PCIA's Coming of Age: 3G Spectrum Conference.

Ercole, R. (2018). 5G in 26 GHz: A technical note on the implications of interference to passive satellite services. PolicyTracker.

Foster, A. (2008). *Spectrum sharing*. Global Symposium for Regulators (GSR). ITU.

Freyens, B. P. (2009). A policy spectrum for spectrum economics. *Information Economics and Policy, 21*(2), 128–144.

Frullone, M. (2007). *A deeper insight in technology and service neutrality*. ITU Workshop on Market Mechanisms for Spectrum Management, Geneva.

Funk, J. L. (1998). Competition between regional standards and the success and failure of firms in the world-wide mobile communication market. *Telecommunications Policy, 22*(4–5), 419–441.

Georgieva, M. (2008). *Co-shaping of governance and technology in the development of mobile telecommunications*. Twente University. Enschede.

Goldman, D. (2010). 4G is a myth (and a confusing mess). *CNNMoney*, Retrieved from https://money.cnn.com/

GSMA. (2019). *Beyond WRC-19: Future spectrum for mobile broadband*. ATU 3rd Working Group meeting, Botswana.

Haug, T. (2002). A commentary on standardization practices: Lessons from the NMT and GSM mobile telephone standards histories. *Telecommunication Policy, 26*, 103–104.

Huurdeman, A. A. (2003). *The Worldwide History of Telecommunications*. John Wiley & Sons.

Indepen and Aegis Systems. (2004). *Costs and benefits of relaxing international frequency harmonisation and radio standards*. stakeholders.ofcom.org.uk

ITU. (1992). WARC-92 concludes after strenuous negotiations. *Telecommunication Journal, 59*(4), 171–175.
ITU. (1997). WRC-97 News. *WRC News*. Retrieved on 30 December, 2014, from http://www.itu.int/
ITU. (1998). The ITU Takes Mobile into the Third Millennium. *Press Release*. Retrieved on 30 April, 2013, from http://www.itu.int/
ITU. (2000a). The dawn of 3G mobile systems. *ITU News*(September). Retrieved on 27 September, 2020, from www.itu.int
ITU. (2000b). WRC-2000 delivers on great expectations. *ITU News*(June). Retrieved on 27 September, 2020, from www.itu.int
ITU. (2000c). WRC-2000 delivers on great expectations. *ITU News*(6).
ITU. (2007). Egypt—Signing of agreement granting mobinil 3G license. *ITU News*. Retrieved on 30 April, 2013, from www.itu.int
ITU. (2008). Development of IMT-advanced: The SMaRT approach. *Press Release*. Retrieved on 30 April, 2013, from www.itu.int
ITU. (2009). IMT-advanced (4G) mobile wireless broadband on the anvil. *Press Release*. Retrieved on 30 April, 2013, from www.itu.int
ITU. (2010a). ITU paves way for next-generation 4G mobile technologies. *Press Release*. Retrieved on 30 April, 2013, from http://www.itu.int/
ITU. (2010b). *ITU world radiocommunication seminar highlights future communication technologies*. Retrieved from www.itu.int
ITU. (2010c). ITU world radiocommunication seminar highlights future communication technologies. *Press Release*. Retrieved on 30 April, 2013 from www.itu.int
ITU. (2012). IMT-advanced standards for mobile broadband communications. *Press Release*. Retrieved on 30 April, 2013, from www.itu.int
ITU. (2017a). *Futuristic mobile technologies foresee IMT for 2020 and beyond*. Retrieved 22-10 from www.itu.int
ITU. (2017b). *ITU agrees on key 5G performance requirements for IMT-2020*. Retrieved from www.itu.int
ITU. (2020). *Overview of ITU's history*. Retrieved on 20 September, 2020 from www.itu.int
ITU-D. (2009). What really is a third generation (3G) mobile technology? *IMT introducing*. Retrieved on 30 November, 2013 from www.itu.int
ITU-D. (2012). *Digital dividend: Insights for spectrum decisions*. Retrieved from www.itu.int
ITU-R. (1997). ITU-R recommendation M.687-2: International mobile telecommunications-2000 (IMT-2000). In *M Series. Mobile, radiodetermination amateur and related satellite services*. ITU.
ITU-R. (2001a). Article 5: Frequency allocations. In *WRC-2000 final acts*. ITU.
ITU-R. (2001b). Resolution 224: Frequency bands for the terrestrial component of IMT-2000 below 1 GHz. In *Provisional Final Acts—World Radiocommunication Conference (WRC-2000)*. ITU.
ITU-R. (2001c). WRC-2001 resolution 223: Additional frequency bands identified for IMT. In *Provisional Final Acts—World Radiocommunication Conference (WRC-2000)*. ITU.
ITU-R. (2003a). *Deployment of IMT-2000 systems*. ITU.
ITU-R. (2003b). *Deployment of IMT-2000 systems handbook*. ITU.
ITU-R. (2007a). ITU-R recommendation M.1822: Framework for services supported by IMT. In *M Series. Mobile, radiodetermination amateur and related satellite services*. ITU.
ITU-R. (2007b). *Resolution ITU-R 56: Naming for international mobile telecommunications* radiocommunication assembly (RA-07), Retrieved from www.itu.int
ITU-R. (2007c). *Resolution ITU-R 57: Principles for the process of development of IMT-advanced* radiocommunication assembly (RA-07), Retrieved from www.itu.int
ITU-R. (2007d). WRC-07 resolution 223: Additional frequency bands identified for IMT. In *Provisional Final Acts—World Radiocommunication Conference (WRC-07)*. ITU.
ITU-R. (2008). Article 5: Frequency allocations. In *Radio regulations* (Vol. 1). ITU.

ITU-R. (2009). ITU-R recommendation M.1457-8: Detailed specifications of the radio interfaces of international mobile telecommunications-2000 (IMT-2000). In *M Series. Mobile, radiodetermination amateur and related satellite services.* ITU.

ITU-R. (2010). *ITU-R report SM.2093–1: Guidance on the regulatory framework for national spectrum management*. ITU.

ITU-R. (2012a). Article 5: Frequency allocations. In *Radio regulations: 2012 Edition* (Vol. 1). ITU.

ITU-R. (2012b). ITU-R recommendation M.2012: Detailed specifications of the terrestrial radio interfaces of international mobile telecommunications advanced (IMT-Advanced). In *M Series. Mobile, radiodetermination amateur and related satellite services.* ITU.

ITU-R. (2012c). WRC-12 resolution 232. Use of the frequency 694–790 MHz by the mobile, except aeronautical mobile, service in region 1 and related studies. In *Provisional Final Acts—World Radiocommunication Conference (WRC-12).* ITU.

ITU-R. (2013a). Draft revision 1 of ITU-R recommendation M.2012: detailed specifications of the terrestrial radio interfaces of international mobile telecommunications advanced (IMT-Advanced). In *M Series. Mobile, radiodetermination amateur and related satellite services.* ITU.

ITU-R. (2013b). ITU-R Recommendation M.1457-11: Detailed specifications of the radio interfaces of international mobile telecommunications-2000 (IMT-2000). In *M Series. Mobile, radiodetermination amateur and related satellite services.* ITU.

ITU-R. (2013c). M.2047: Detailed specifications of the satellite radio interfaces of international mobile telecommunications-advanced (IMT-Advanced). In *Recommendations: M-Series.* ITU.

ITU-R. (2014). Recommendation M.1850: Detailed specifications of the radio interfaces for the satellite component of international mobile telecommunications-2000 (IMT-2000). In *Recommendations: M-Series.* ITU.

ITU-R. (2015a). Article 5: Frequency allocations. In *Radio regulations* (Vol. 1). ITU.

ITU-R. (2015b). M.1036-5 (10/2015): Frequency arrangements for implementation of the terrestrial component of international mobile telecommunications (IMT) in the bands identified for IMT in the Radio Regulations. ITU.

ITU-R. (2015c). M.2083: IMT vision—"framework and overall objectives of the future development of IMT for 2020 and beyond". In *Recommendations: M-Series.* ITU.

ITU-R. (2015d). WRC-15 Resolution COM4/6 (WRC-15), Review of the spectrum use of the frequency band 470–960 MHz in Region 1. In *Provisional Final Acts - World Radiocommunication Conference (WRC-15).* ITU.

ITU-R. (2016a). Circular Letter 5/LCCE/59: To administrations of member states of the ITU, radiocommunication sector members, ITU-R associates participating in the work of radiocommunication study group 5 and ITU academia. Retrieved from www.itu.int

ITU-R. (2016b). Invitation for submission of proposals for candidate radio interface technologies for the terrestrial component of the radio interface(s) for IMT-2020 and invitation to participate in their subsequent evaluation. Retrieved from www.itu.int

ITU-R. (2016c). Resolution 238: Studies on frequency-related matters for international mobile telecommunications identification including possible additional allocations to the mobile services on a primary basis in portion(s) of the frequency range between 24.25 and 86 GHz for the future development of international mobile telecommunications for 2020 and beyond. In *Radio regulations.* ITU.

ITU-R. (2019a). Addendum 5 to circular letter 5/LCCE/59 to administrations of member states of the ITU, radiocommunication sector members, ITU-R Associates participating in the work of radiocommunication study group 5 and ITU Academia. Retrieved from www.itu.int

ITU-R. (2019b). Recommendation M.1036–6: Frequency arrangements for implementation of the terrestrial component of International Mobile Telecommunications (IMT) in the bands identified for IMT in the Radio Regulations In *ITU-R Recommendations: M-Series.* ITU.

ITU-R. (2020a). Addendum 6 to circular letter 5/LCCE/59: Completion of Step 4 of the IMT-2020 process for the evaluation of IMT-2020 candidate technology submissions. Retrieved from www.itu.int

ITU-R. (2020b). Addendum 7 to circular letter 5/LCCE/59: Announcement of the IMT-2020 radio interface technologies resulting from the completion of Steps 5 to 7 of the IMT-2020 process for the evaluation of IMT-2020 candidate technology submissions and decision on the qualifying RIT/SRIT technologies for IMT-2020 that advance to Step 8. Retrieved from www.itu.int

ITU-R. (2021a). Addendum 8 to circular letter 5/LCCE/59: information on completion of the first release of the detailed specifications of the radio interfaces for the terrestrial component of international mobile telecommunications-2020 (IMT-2020) in Recommendation ITU-R M.2150 and future plans for 'IMT—2020 and beyond'. www.itu.int

ITU-R. (2021b). ITU-R recommendation M.2150: Detailed specifications of the terrestrial radio interfaces of international mobile telecommunications-2020 (IMT-2020). In *ITU-R Recommendations M-Series*. ITU.

Jho, W. (2007). Global political economy of technology standardization: A case of the korean mobile telecommunications market. *Telecommunications Policy, 31*(2).

Kamecke, U., & Korber, T. (2006). Technological neutrality in the EC regulatory framework for electronic communications: A good principle widely misunderstood. *E.C.L.R.*(5), 124–138.

Leite, F., Engelman, R., Kodama, S., Mennenga, H., & Towaij, S. (1997). Regulatory considerations relating to IMT-2000. *IEEE Personal Communications, 4*(4), 14–19.

London Economics. (2008). *Economic impacts of increased flexibility and liberalisation in European spectrum management: Report for a group of european communications sector companies.* Retrieved from http://londoneconomics.co.uk/

Lovells, H. (2014). Technology neutrality in internet, telecoms and data protection regulation. *Global Media and Communications, Autumn*(14), 19–23.

Manner, J. A. (2003). *Spectrum wars: The policy and technology debate.* Artech House.

Marti, M. R. (2019). Europe hoping to overcome C-band satellite interference. PolicyTracker.com.

Newlands, M. (2009). World's biggest mobile market rejects technology neutrality. *PolicyTracker*. Retrieved on 19 March, 2012 from www.policytracker.com

NTRA of Egypt. (2006a). *Mobinil postpones promoting EDGE services.* Retrieved on 30 October, 2014 from www.ntra.gov.eg

NTRA of Egypt. (2006b). *Reality about EDGE problem.* Retrieved on 30 October, 2014 from www.ntra.gov.eg

Osseiran, A. (2013). *The 5G mobile and wireless communications system.* ETSI Future Mobile Summit.

Pogorel, G. (2007). Opinion: The nine regimes of spectrum management. *PolicyTracker*. Retrieved on 28 July, 2007 from www.policytracker.com

Rancy, F. O. (2017). Standards and spectrum for international mobile telecommunications. *ITU News*(February). Retrieved on 27 September, 2020 from www.itu.int

Russian Federation. (2018). *Proposals for further consideration in the framework of Recommendation ITU—R M. 1036 the frequency bands not identified for IMT* WP 5D Meeting.

SADC-EACO. (2019). *Agenda item 10, input contribution to ATU*. ATU Working Group 3rd Meeting, Botsawana.

Saugstrup, D., & Henten, A. (2006a). 3G Standards: the battle between WCDMA and CDMA2000. *Info, 8*(4).

Saugstrup, D., & Henten, A. (2006). 3G Standards: The battle between WCDMA and CDMA2000. *Info, 8*(4), 10–20.

Savage, J. (1989). *The politics of international telecommunications regulation.* Westview Press.

Sims, M. (2005). Move Towards Technology Neutrality. *PolicyTracker.com.* Retrieved on 30 April, 2012, from www.policytracker.com

Sims, M. (2006a). Commission admits defeat. *PolicyTracker.com.* Retrieved on 30 April, 2012, from www.policytracker.com

Sims, M. (2006b). Europe divided, Asia takes the lead. *PolicyTracker.com.* Retrieved on 30 April, 2012 from www.policytracker.com

Sims, M. (2007a). CDMA2000 gets special treatment in Hong Kong. *PolicyTracker.com.* Retrieved on 30 April, 2012 from www.policytracker.com

Sims, M. (2007b). Geneva ITU meetings increase flexibility for mobile. *PolicyTracker*. Retrieved on 19 March, 2012 from https://www.policytracker.com

Standeford, D. (2018). Debate over interference limits in 26 GHz heats up. PolicyTracker.com. Retrieved on 19 March, 2019 from https://www.policytracker.com

U.S. Congress—Office of Technology Assessment. (1991). *The 1992 World Administrative Radio Conference: Issues for U.S. International Spectrum Policy.* Washington D.C.

U.S. Congress Office of Technology Assessment. (1993). *The 1992 World Administrative Radio Conference: Technology and Policy Implications.*

Whittaker, M. (2002). *True Technology Neutral Spectrum Licences.* Retrieved from www.futurepace.com.au

WIMAXForum. (2007). WiMAX and IMT-2000. Retrieved on 30 April, from http://www.wimaxforum.org

WP 5D Chairman. (2007a). *Chapter 01—Working Party 5D Chairman's Report* Meeting of Working Party 5D (Kyoto, 23 to 31 May 2007), Geneva. Retrieved from www.itu.int

WP 5D Chairman. (2007b). *Chapter 01—Working Party 5D Chairman's Report* Meeting of Working Party 5D (Seoul, 28 to 31 August 2007), Geneva. Retrieved from www.itu.int

WP 5D Chairman. (2009). *Chapter 01—Working Party 5D Chairman's Report* Meeting of Working Party 5D (Dresden, 14 to 21 October 2009), Geneva. Retrieved from www.itu.int

Xia, J. (2011). The third-generation-mobile (3G) policy and deployment in China: Current status, challenges, and prospects. *Telecommunications Policy, 35*(1), 51–63.

Youell, T. (2018). US regulator rejects broadcasters' bid to undo IMT identification at 470–608 MHz. *PolicyTracker*. Retrieved on September 27, 2020 from https://www.policytracker.com

Zacher, M. W. (1996). *Governing global networks: International regimes for transportation and communications.* Cambridge University Press.

Zelmer, D. (2002). GSM goes to North America. In F. Hillebrand (Ed.), *GSM and UMTS: The creation of global mobile communication* (Vol. 105–114). Wiley.

Ziegler, C. (2011a). 2G, 3G, 4G, and everything in between: an Engadget wireless primer. *Engadget*. Retrieved on August 1, 2021 from https://www.engadget.com/

Ziegler, C. (2011b). AT&T: both HSPA+ and LTE are '4G,' 20 such devices planned for this year. *Engadget*. Retrieved on August 1, 2021 from https://www.engadget.com/

Chapter 6
Spectrum Usage Rights

If we knew all the situations in which property rights or commons were more efficient, we could simply mandate that world. But we do not know that, nor can we.

Werbach (2004)

6.1 Introduction

Determining the type of spectrum usage rights is an important component of spectrum management. These rights can be exclusive, where spectrum users have individual access to the spectrum, or collective, where there is common access to spectrum without exclusive property rights. A third type of rights is easement, which entails one entity allowing other users access to the spectrum that is owns or has licensed as long as they do not cause harmful interference. Easements are possible in two forms: overlay (opportunistic) and underlay. Overlay devices access the spectrum at the geographical, time or frequency gaps of the licensed user's transmission as long as it does not cause harmful interference (e.g. TVWS). Underlay access implies that a secondary user will transmit at low power levels, within the noise floor of licensed spectrum (e.g. UWB).

At the international level, the concept of spectrum usage right is not explicitly mentioned or defined in the RR. However, the rights of spectrum access are organized through the radiocommunication service allocation framework that prioritizes access to the spectrum band among the different services allocated within it. In addition, there are several provisions within the RR that outline spectrum usage rights in terms of different characteristics including maximum transmitted power, power flux-density and define coordination procedures. In all cases, the main principle of the RR is that radio stations must not cause harmful interference to other stations that operate in accordance with the provisions of these regulations.

With this in mind, in this chapter, we highlight the elements within the international spectrum management that are related to spectrum usage rights. Thus, the chapter starts with exploring the spectrum usage right concept in general. The chapter then

M. A. El-Moghazi and J. Whalley, *The International Radio Regulations*,
https://doi.org/10.1007/978-3-030-88571-7_6

explores the concept within the context of the RR before focusing on five cases of spectrum usage right and analyses them in terms of how in each of them spectrum usage rights are defined and how they fit within the RR framework.

6.2 Usage Rights

In order to understand the concept of property rights in spectrum, it is firstly useful to understand what is meant by property or usage rights. Demsetz (1967) points out that property rights were developed to internalize externalities when the gains of internalization become larger than the cost of internalization. In other words, property rights are developed only when their benefits exceed their expenses. Hazlett (2006) defined four categories of property in terms of their access rights:

- open access property in which there is no restrictions on usage;
- state property where access rules are determined by the state;
- common property where access rules are determined by owners;
- private property where access rules are determined by a single owner.

In Chapter Two, we explored the three types of spectrum usage rights: exclusive property rights, exclusive property rights with easement or collective (non-exclusive). Generally speaking, spectrum usage rights entail specifying technical restrictions to protect neighbouring users from harmful interference assuming certain application and technology for both the interfering party and the victim (Ofcom, 2006). Spectrum usage rights could also be understood in terms of rights and duties where the main user has the right to access spectrum while other users have the duty not to access this spectrum (Hohfeld, 1913).

The first attempt to design spectrum property rights was in 1969 when Vany et al. (1969) proposed a spectrum property right package called 'TAS', which stands for *time* (the time during which transmission occur), *area* (the geographic area over which the radio waves spread) and *spectrum* (the range of frequencies). However, the proposal was abandoned by the FCC. Subsequently, several scholars have attempted to develop alternatives designs for spectrum property rights (Cave & Webb, 2003; Vries & Sieh, 2012). For instance, one approach was to define out-of-band and in-band power. The first sets out the spillover interference into the neighbouring bands, while the second determines the maximum transmitted power in its own band. Setting values for both elements would enable the users to determine which technology and usage would fit within the limits of the spectrum property rights (Cave & Webb, 2003).

Vries and Sieh (2012) proposed a reception-oriented radio property rights model that is based on three phases. The first phase defines the operating rights and harmful interference through outlining transmission permissions and reception protections limits. The second phase facilitates transactions through limiting the number of parties involved in a negotiation, only altering the rights in a license at renewal and using a registry to record current parameter values. The third phase makes rights enforcement efficient, cheap and predictable through enabling the direct enforcement of rights, separating rulemaking from adjudication and defining remedies up-front.

The practice of developing spectrum property rights has varied around the world with no example of significant success occurring. For instance, Ofcom in the UK developed a model called 'spectrum usage rights' (SUR). SUR specifies the maximum level of interference that can be caused, rather than the power that can be transmitted (Ofcom, 2008). There were, however, some difficulties associated with SUR (Eurostrategies & LS-Telecom, 2007), and accordingly, the approach has not been widely applied. In addition, the FCC studied an approach to defining spectrum property rights called 'interference temperature'. In this approach, transmission is permissible as long as the resulting interference at any unintended receiver does not exceed a certain level (Evci & Fino, 2005). The approach was quite complex contributing to the decision of the FCC in 2007 to abandon it as it was impractical and may increase interference levels (Weiser & Hatfield, 2008).

Guatemala applied in 1996 a private property rights approach to spectrum management. In such an approach, spectrum rights are called 'titulo de usufructo de frecuencia' (TUF), and they define ownership by specifying parameters such as the band or frequency ranges, hours of operation, geographical coverage area, maximum effective radiated power and the maximum field strength or signal strength on the border of the coverage area (Hazlett & Muñoz, 2006). Another approach was adopted in Australia called 'space centric management'. This is based on defining the maximum transmit power at the antenna rather than maximum field strength away from the antenna (Whittaker, 2007).

While there is a limited literature on the assessment of the influence of these approaches, the concept of property rights has helped solve some interference situations in practice without referring explicitly to the concept but rather through implicitly adopting it (e.g. interference from 4G base stations to radars systems and TV receivers in France) (Deffains, 2013). Furthermore, it seems that countries tend to assign spectrum usage rights that are aligned with the RR, and one of the main reasons of revoking such rights is related to binding international agreements (ECC, 2011).

We also pointed out in Chapter Two that easement is a right to use the property of another without possessing it and that exclusive spectrum property rights with easements allows other users to access spectrum that is owned or licensed to someone else without causing harmful interference. The two types of spectrum easements accommodate overlay where devices access the spectrum at the geographical, time or frequency gaps of the licensed user's transmission, whereas underlay is where a secondary user will transmit at low power levels within the noise floor of licensed spectrum (Cave & Webb, 2012).

Within underlay access, spectrum usage rights are determined by calculating the allowed interference power to be received by the licensed service which is dependent on the underlay devices transmit power (Cave & Webb, 2012). Underlay devices could take the form of a short range devices (SRD) which transmit using a small amount of bandwidth and power over short distances or in the form of UWB which can transmit over a broad range of bandwidth with very low power levels (Cave & Webb, 2003). The situation is similar for overlay access with the exception that higher transmitting powers are allowed for overlay devices in the case where there is some distance between these devices and the primary or main user (Cave & Webb, 2012).

Overlay or opportunistic access is promoted by technologies such as cognitive radio system (CRS) that are capable of measuring the radio environment and learning from experience in order to transmit dynamically in the temporal unused frequencies without the need for exclusive allocation. One of the main candidate spectrum bands for CRS operations is TVWS, which refers to the geographically interleaved vacant frequencies in TV spectrum. Overlay access could be organized using a database that contains the maximum transmitted power for overlay users according to their position (Cave & Webb, 2012). Alternatively, QinetiQ (2005) proposed utilizing spectrum monitoring systems to allow the simultaneous use of parts of the spectrum by primary and secondary users.

The third type of spectrum usage rights is collective usage rights, which entails common access to spectrum without having exclusive property rights in the spectrum by a particular entity (Chaduc & Pogorel, 2008). The concept of usage rights in collective access is different where users are divided depending on the amount of interference they cause, requiring manufacturers to adopt fairness protocols including control over power levels, develop a geographical database or sensing the radio environment (Cave & Webb, 2012).

Collective usage rights are argued to have more benefits than underlay or overlay access due to the restrictions placed on the latter (Lehr & Crowcroft, 2005). Moreover, it has been suggested that dividing the spectrum into bands of frequencies was an architecture design for enabling communications and technical approach to interference management (Werbach, 2004). This may not reflect the nature of spectrum. For example, concepts such as 'spread-spectrum' enable multiple users to access the spectrum. This is based on the theory that a signal can either be sent across a narrow channel at high power or spread across a wide channel at lower power (Shannon, 1948). The technological focus was historically on the former. In addition, Benkler (2002) points out that commons will outperform property rights regime because wireless communications capacity is mostly determined by local conditions and that human communications are highly variable within geographically small areas and over time.

6.3 Usage Rights in the RR

The concept of spectrum usage rights is not explicitly mentioned or defined in the RR. However, the rights of spectrum access are organized through the radiocommunication service allocation framework that prioritizes access to the spectrum among the different services allocated within the band. Moreover, we can also say that the RR framework has been based on sharing since its establishment. The notion of allocating the radio spectrum into different radiocommunication services started with the first International Radiotelegraph Convention of 1906 that divided spectrum into public correspondence, military and naval bands (Savage, 1989). The service priority concept was not introduced until the RR of 1959 that formally established the different service priorities: permitted and secondary services. Permitted services have the same priority as primary services except in the case of preparation plans where the former has less priority than the latter. Secondary services cannot cause interference to both primary and permitted services. At the RR of 1996, the concept of permitted service was abandoned while retaining primary and secondary services (Ard-Paru, 2013).

In Chapter Three, we explained that the stations of a secondary service cannot cause harmful interference to stations of primary services and cannot claim protection from harmful interference caused by stations of a primary service (ITU-R, 2020h). The operation of secondary services is usually associated with conditions to ensure the protection of primary services. One example is in the 5351.5–5366.5 kHz band where the amateur service should not exceed a maximum radiated power of 15 W (e.i.r.p.) according to Footnote 5.133B so that it can coexist with the fixed and mobile services operating in the band on a primary basis (ITU-R, 2020h). In other cases, there is an additional allocation to a secondary service in particular countries with some conditions being imposed. For instance, Footnote 5.181 provides for additional secondary allocation—in a few countries this occurs in the 74.8–75.2 MHz band with the mobile service subject to agreement obtained under No. 9.21.

In other cases, the RR accommodate the complex situations that exist between primary and secondary services. For instance, the 223–230 MHz band is allocated to broadcasting services on a primary basis and to the fixed and mobile services on a secondary basis. However, Footnote 5.246 implies alternative allocation in a few countries where the 223–230 MHz band is allocated to the broadcasting and land mobile services on a primary basis as long as that in the preparation of frequency plans the broadcasting service will have the first choice of frequencies (ITU-R, 2020h).

It is not common in the RR table of service allocation to find an exclusive allocation to only one service. For instance, in Region 1, in the 24.25–24.45 GHz band, there is only allocation to the fixed service on a primary basis (ITU-R, 2020h). Instead, the RR often accommodate two or more primary services in the same band. Of course, this does not entail that these services can operate simultaneously without coordination. Thus, coordination between services is essential, even among those primary services that are allocated in the same band. For instance, in the RR of 2016, several frequency bands in the 275–1000 GHz range were identified for passive service applications, and active services were urged to take all practicable steps to protect these passive services from harmful interference. Following WRC-19, in several bands in the 275–1000 GHz range, fixed and land mobile service applications were allowed to operate to ensure the protection of Earth exploration-satellite service (passive) applications (ITU-R, 2020h).

The RR allow, in some cases, for the operation of two primary services with equal access in the same band. For instance, Article 21 specifies some rules for the choice of sites and frequencies for terrestrial and space services sharing frequency bands above 1 GHz in terms of geographical separation between earth stations and terrestrial stations, power limits for terrestrial stations and the limits of power flux-density from space stations (ITU-R, 2020f). Determining the limits to ensure two primary services can coexist is not an easy task due to the continuous development of these services. For instance, Article 21.5 states some limits to the power delivered by a transmitter to the antenna of a station in the fixed or mobile services to ensure sharing between terrestrial and space services. During WRC-19, it was deemed necessary to study the applicability of these limits to IMT stations, which use an antenna that consists of an array of active elements (ITU-R, 2019a). Another important component of the RR that address sharing between passive services and active services operating in adjacent bands is Resolution 750. This resolution determines the limits of unwanted emissions of active services (ITU-R, 2020i).

As we have demonstrated in Sect. 6.2, the RR framework of service allocation accommodates sharing. Accordingly, several proposed models of sharing were designed to fit within such a framework. For instance, ASA (see Chap. 2) is aligned with the RR where ASA is to be deployed in bands allocated to mobile service and identified for IMT on a non-interference basis. These bands include the 2.3 GHz and 3.8 GHz, where ASA would share with military applications and satellite services, respectively (ECC, 2013). Similarly, the LSA concept does not operate on a non-interference or secondary basis as it has a predictable QOS for the main service and the LSA licensee (ECC, 2013). LSA devices are perceived to operate according to the RR service allocation (RSPG, 2011).

Another proposed concept of opportunistic access is ‘pluralistic licensing’ where opportunistic secondary spectrum access can cause interference with primary services whose parameters and rules are known. It has been suggested that such a concept could be used in bands where there are less critical services and secondary use would have more flexibility (Holland et al., 2012). To this end, it is worth highlighting that the RR include a term called ‘common use’ where frequencies should not be notified to the ITU-R to be registered in the MIFR (ITU-R, 2020c). However, they are published in the ITU-R International List of Frequencies (IFL) that contains particulars of frequency assignments recorded in the MIFR (ITU-R, 2020d). One example of ‘common use’ is shown in Article 52.221.3 as follows: *‘The carrier frequencies 4125 kHz, 6215 kHz, 8291 kHz, 12,290 kHz and 16,420 kHz are also authorized for common use by coast and ship stations for single-sideband radiotelephony on a simplex basis for distress and safety traffic’* (ITU-R, 2020g). Other examples of global common use frequencies can be found in Appendix 27 for the aeronautical mobile (R) service (ITU-R, 2020b) and in Appendix 18 for VHF maritime mobile (ITU-R, 2020a).

6.4 Examples of Spectrum Uses

In the previous section, we demonstrated that different forms of spectrum usage rights exist within the RR framework, each with their own format. Exclusive spectrum usage right could be considered as a case of exclusive primary service within a band. Secondly, the spectrum easements approach was shown, through an example of coexisting primary and secondary services, not to cause interference or result in a claim for protection against interference from primary services. The second form of spectrum easements could be considered as the operation of non-interference basis (NIB) according to RR Article 4.4. The third and fourth forms are the sharing between coprimary services and between cosecondary services in the same band, respectively. Finally, the concept of common use is similar to collective usage rights. In the following subsections, we will draw on these findings to analyse five different cases of spectrum uses. Through doing so, we shall explore how they fit within the RR framework.

6.4.1 Industrial Scientific Medical Applications

It is important to trace the origin of Industrial Scientific Medical (ISM) as most Wi-Fi devices today operate in bands designated for ISM. The start can be traced back to when there were exclusive allocations to stations of low power in the RR of 1932 and 1938 (ITU-R, 1932, 1938). These allocations merged with ISM during the RR of 1947 (Ard-Paru, 2013). More specifically, the WRC of 1947 and 1959 allocated several frequency bands to ISM applications excluding applications in the field of telecommunications (ITU-R, 1979). Such an allocation required that the radiation from equipment used for ISM applications was minimal and that outside the bands designated for ISM equipment, radiation levels are such they do not cause harmful interference to radiocommunication services (ITU-R, 2008b).

Within the RR, Article 1.15 of defines ISM as '*industrial, scientific and medical (ISM) applications (of radio frequency energy): Operation of equipment or appliances designed to generate and use locally radio frequency energy for industrial, scientific, medical, domestic or similar purposes, excluding applications in the field of telecommunications*'. There are two main footnotes that identify bands for ISM. The first is Footnote 5.138 which designates the 6765–6795 kHz, 433.05–434.79 MHz (in most region 1 countries), 61–61.5 GHz, 122–123 GHz and 244–246 GHz bands for ISM application. These bands can be utilized as long as special authorization by the relevant administration is granted and that other administrations whose radiocommunication services might be affected also agree. The latest ITU-R recommendations must be adhered to.

On the other hand, Footnote 5.150 contains significantly different conditions for ISM operation. In particular, it designated the 13,553–13,567 kHz, 26,957–27,283 kHz, 40.66–40.70 MHz, 902–928 MHz (in Region 2), 2400–2500 MHz, 5725–5875 MHz and 24–24.25 GHz bands for ISM application and states that radiocommunication services operating within these bands must accept harmful interference which may be caused by these applications. However, Article 15.13 instructs that '*Administrations shall take all practicable and necessary steps to ensure that radiation from equipment used for industrial, scientific and medical applications is minimal and that, outside the bands designated for use by this equipment, radiation from such equipment is at a level that does not cause harmful interference to a radiocommunication service and, in particular, to a radionavigation or any other safety service operating in accordance with the provisions of these Regulations*' (ITU-R, 2020e).

In other words, ISM applications operating in the second group of bands, including those frequencies where Wi-Fi operates such as 2.4–2.5 GHz, could be considered superior to those radiocommunication services operating in these bands. The latter cannot complain harmful interference, though there is a caveat in Article 15.13—the radiation of these applications is minimal within the ISM bands.

Agenda Item 8.1.1 Issue A of WRC-12 addressed an important issue: the protection of radiocommunication services against interference from ISM equipment within and outside the frequency bands designated for ISM applications according to the RR footnotes No. 5.138 and No. 5.150 (ITU-R, 2007a). WRC-12 concluded that limits still needed to be imposed on radiation from ISM equipment, within and outside the frequency bands designated in the RR (Al-Rashedi, 2014).

6.4.2 Short-Range Devices

The origin of short-range devices (SRD) can be traced back in 1938 when the FCC exempted several devices (e.g. baby monitors) from licensing while acknowledging the role of the RR. In other words, the USA set a precedent of a country departing from the RR as long as interference is not caused to stations in other countries (Horvitz, 2013).

Generally speaking, SRD are not defined in the RR. They are, therefore, not regarded as a radio service and do not have primary or secondary status. Accordingly, they operate as outlined by RR Article 4.4 on a non-interference non-protection basis (Anker & Lemstra, 2011). In other words, SRD devices have a lower status with less rights than primary and secondary services and can be called a 'permitted use' (Forge et al., 2012). However, SRD are defined in one of the ITU-R reports as devices that have a low capability of causing interference to other radio equipment and are permitted to operate on a non-interference and non-protected basis (ITU-R, 2019b). This report also clarifies that SRD usually operate within bands designated to ISM applications and where this happens they must accept the harmful interference that may be caused by these applications.

Agenda Item 1.22 of WRC-12 examined the effect of emissions from SRDs on radiocommunication services. Resolution 953 stated that to adequately protect radiocommunication services, further studies of the emissions from SRDs were required (ITU-R, 2007c). The resolution recognized that SRD systems may operate on a non-interference and non-protected basis in ISM and non-ISM bands and that they can also operate under a particular service. During the study period, the idea of having a formal definition of SRD application and conditions where they can operate was discussed. However, concern was expressed that SRDs may no longer operate on a non-interference basis and may claim protection from existing radiocommunication services (ITU-R, 2011a). The studies concluded that there was no need to modify the RR or to continue with studies of SRD harmonization (Al-Rashedi, 2014).

ITU-R Recommendation SM.1896 includes frequency ranges for the harmonization of SRD. In particular, the recommendation identifies a number of frequencies for global harmonization (e.g. 3155–3400 kHz, 3.7–4.8 GHz and 244–246 GHz) as well as frequency ranges for use in possible regional harmonization (e.g. 312–315 MHz, 875–960 MHz and 6–9 GHz). The recommendation emphasized that SRDs are not ISM applications and that UWB applications are a subset of SRD applications and the use of radio spectrum on a non-interference and non-protected basis with respect to radiocommunication services (ITU-R, 2018). RFID devices are also a subset of SRD and are often deployed in the ISM bands. Harmonization of RFID would reduce the potential for their harmful interference with radiocommunication services (ITU, 2012b).

6.4.3 Radio Local Access Network (RLAN)

The origin of RLAN goes back to 1985 when the FCC decided to open up three ISM spectrum bands (915 MHz, 2.4 GHz and 5.8 GHz) to wireless local area network (WLAN) (Lemstra et al., 2011a). The decision led to a similar approach being adopted in Europe for the 2.4 GHz band (Lemstra et al., 2011).

There was a perception during the 1990s that the operation of Wi-Fi suffered from interference from SRD devices operating in the 2.4 GHz band and that there was a need for additional spectrum in the 5 GHz band to achieve higher data rates (Negus & Petrick, 2009). Meanwhile, in Europe, the 5150–5350 MHz bands were recommended to be used for what was called 'HIPERLAN' equipment on a non-protected and a non-interference basis (ERC, 1992). The industry in the USA was motivated by this decision to call for, in 1995, a similar allocation in the 5 GHz band for WLAN. Consequently, the FCC allocated the 5150–5350 MHz and 5725–5825 MHz spectrum bands for the operation of the unlicensed national information infrastructure (U-NII) devices (Anker & Lemstra, 2011). For both cases, WLAN devices were designated by regulators on a non-protection, non-interference basis with low power (Anker & Lemstra, 2011).

In 2000, the European countries proposed that WRC-2000 would have an agenda item in WRC-03 to consider additional spectrum allocation to mobile services to obtain global harmonized frequency allocations for RLAN (CEPT, 2000). Their argument was that the global mobile service allocation would give RLANs an appropriate ITU allocation status. Moreover, the ISM band at 5 GHz was not preferred due to the expected significant increase in the use of other applications in this band and the high quality of service requirement for RLANs (CEPT, 2000). However, it was recognized that such allocation would require sharing with other systems (e.g. radar systems) and therefore the ITU studied the technical restrictions necessary to permit such sharing (Horne, 2003).

Eventually, WRC-03 decided to allocate the 5150–5350 MHz and 5470–5725 MHz bands on a primary basis to mobile services for the implementation of wireless access systems including RLANs (ITU-R, 2003b). WRC-03 also agreed to allow sharing of the 5 GHz bands between radiolocation service (radar systems) and mobile service (RLAN) on a primary basis using dynamic frequency selection (DFS) and transmit power control (TPC) (ITU-R,). Horne (2003) explains that it was unusual for the ITU to determine technical restrictions through behaviour-based mechanisms such as DFS. It was the first time that mandated sharing conditions associated with an allocation had occurred. Furthermore, the R-LAN mobile service allocation in the 5 GHz band is considered a rare example of where a primary radiocommunication service allocation was assigned to a usage that could be considered to operate on a non-interference basis. Such primary allocation has two advantages: firstly, providing certainty for RLAN systems in the 5 GHz and secondly providing protection for radar system operating in the band (Anker & Lemstra, 2011).

One important observation is that in developing the cognitive feature for accessing the spectrum in the 5 GHz by the RLAN, the characteristics of the operating systems in the band were, at that time, known (Zhao et al., 2007). Therefore, while it may not have been possible to have rules for service sharing and opportunistic access in all spectrum bands, for a specific band and particular operating system, this could be developed within the ITU-R. It has been argued that one of the main reasons for the success of Wi-Fi in the ISM bands was because of the usage of spread spectrum techniques. These are resistant to interference, which is important given that, in these bands, radiocommunication services must accept harmful interference which may be caused by ISM applications (Hayes & Lemstra, 2008). It should be noted that there is a key difference between the USA and Europe when it comes to utilizing ISM bands. While the USA has more contiguous frequencies in the 900 MHz band, Europe has fewer frequencies in the 868 and 433 MHz bands. In addition, the USA allows higher power limits for utilizing the ISM bands (Benkler, 2012).

6.4.4 Cognitive Radio System

It is important to address cognitive radio systems (CRS) in the context of the RR so that it can be examined whether or not it fits within the international framework for spectrum management. ITU-R started to focus on CRS during WRC-07 when there were calls to consider spectrum requirements and global allocations to support CRS and SDR. One of these proposals was from CEPT, which was motivated by the concept of a cognition-supporting pilot channel (CPC) to enable a terminal to select a network (Billquist, 2008). As a result, WRC-07 issued Resolution 956 that invited the ITU-R to study measures such as the need for a worldwide harmonized CPC or to create a database that can assist in the determination of local spectrum usage (ITU-R, 2007d). In addition, it was decided that the ITU-R would study whether there was a need for regulatory measures related to the application of CRS technologies and software-defined radio (ITU-R, 2007b).

One of the main outcomes of the ITU-R following WRC-19 was agreeing on definitions for CRS and SDR (ITU-R, 2009b). CRS was defined as *'[A] radio system employing technology that allows the system to obtain knowledge of its operational and geographical environment, established policies and its internal state; to dynamically and autonomously adjust its operational parameters and protocols according to its obtained knowledge in order to achieve predefined objectives; and to learn from the results obtained'* (ITU-R, 2009b). In contrast, SDR was defined as: '*A radio transmitter and/or receiver employing a technology that allows the RF operating parameters including, but not limited to, frequency range, modulation type, or output power to be set or altered by software, excluding changes to operating parameters which occur during the normal pre-installed and predetermined operation of a radio according to a system specification or standard'* (ITU-R, 2009b).

Following WRC-07, there were proposals to require band-by-band studies before using CRS in any band (Billquist, 2010). The interest in CPC declined following WRC-07 (Ofcom, 2012), while during WRC-12 some concerns were expressed regarding interference between CRS and space, passive and safety services (RCC, 2011). Countering this, others argued that national regulators can set operating parameters for CRS devices so that they did not cause any interference (CEPT, 2011; CITEL, 2012). Several regional organizations called for the development of a WRC resolution to provide a framework to guide the study of CRS as well as how the use of CRS should be administered (APT, 2011; ASMG, 2011; ATU, 2011).

Eventually, WRC-12 did not decide on any particular measure with regard to CRS, with CRS being recognized as technologies and not radiocommunication services. It was also agreed that the examination of the implementation and use of CRS in radiocommunication services should continue without the need for consideration at subsequent WRCs (ITU-R, 2012c). Any radio system implementing CRS technology should operate in accordance with the RR and that the use of CRS does not exempt administrators from their obligations (ITU-R, 2012d).

Having said this, it was argued that as no explicit regulation in the RR forbids CRS devices, they can operate within the current allocation framework whether on a secondary or primary basis while utilizing a suitable service (such as mobile). Alternatively, CRS can operate on a non-interference non-protection basis with respect to stations operating in accordance with RR Article 4.4. This could only be applied to cross-border cases as countries are sovereign within their territories (Anker, 2010). Furthermore, the WRC-12 decision on A.I. 1.19 could be considered as indirect support for the implementation of CRS because the alternative decision, to conduct band-by-band studies before using CRS, would delay the deployment of CRS (El-Moghazi et al., 2012).

There are, however, several forces that could deter the deployment of CRS. For instance, countries usually register their frequency assignments that may have international implications in the MIFR to obtain appropriate recognition (ITU-R, 2012a). Such registration should include fixed parameters such as frequency, bandwidth and transmitted power (ITU-R, 2008a). However, CRS dynamically adjusts its operational parameters according to the operational and geographical environment (ITU-R, 2009a). Therefore, CRS stations may not be registered in the MIFR. There were also some concerns regarding the use of CRS in bands shared with other services that have specific technical or operational characteristics, such as space services (space-to-Earth) and radiodetermination services (ITU-R, 2012b).

6.4.5 TV White Spaces

TVWS has attracted a lot of attention within the ITU in recent years as a promising way to provide broadband services in rural areas while not utilizing exclusive spectrum. TVWS refers to those geographical interleaved vacant frequencies in broadcasting spectrum. Those frequencies were allocated to broadcasting services but were not used in a particular area or frequency because of the need for a spectrum guard band and geographical separation between TV channels to avoid interference (Freyens & Loney, 2011a, b).

Several countries expressed their reservations about TVWS. For instance, the UAE announced that there is a challenge when deploying TVWS devices because the spectrum available for TVWS is decreasing (Almarzooqi, 2014). Egypt also expressed their views on TVWS in 2014 (NTRA of Egypt, 2014), recommending that commercial TVWS deployment should be postponed until the finalization of the DSO process and following the WRC-15 to provide certainty for both operators and end-users. Even in countries that promoted TVWS such as the USA, decisions were taken that may have negatively influenced TVWS deployment. For instance, in 2016, the FCC started its incentive auction of 600 MHz that was expected to evacuate 126 MHz for mobile operators (Yahoo Finance, 2016).

Within the ITU, there has been a work regarding the definition of TVWS which was defined in one ITU-R report as '*A portion of spectrum in a band allocated to the broadcasting service and used for television broadcasting that is identified by an administration as available for wireless communication at a given time in a given geographical area on a non-interfering and non-protected basis with regard to other services with a higher priority on a national basis*' (ITU-R, 2011c).

Following WRC-12, there have been some work undertaken in ITU-R Working Party (WP) 1B to produce a report on TVWS (Chairman of Working Party 1B, 2013). Although TVWS has been defined in one of the ITU-R reports (ITU-R, 2011c), the TVWS definition is not yet been formally defined by the ITU-R. This is similar to the cases of CRS and SDR (ITU-R, 2013). Furthermore, work within the ITU-R focused more in 2014 on DSA in general rather than TVWS in particular (Chairman of Working Party 1B, 2014), and then, the focus changed once again in 2015 towards CRS (Chairman of Working Party 1B, 2015).

WRC-15 contained both good and bad news for TVWS. On the one hand, in some countries in Regions 2 and 3, the 470–694 MHz band was identified for IMT. This reduced the possibilities for TVWS in these countries. On the other hand, in Region 1 countries, it was decided to maintain the broadcasting service allocation status in the previous band for terrestrial broadcasting until WRC-23 (ITU-R, 2015). With regard to the service status of TVWS, one view from CEPT was that they should operate on non-interference basis (Newlands, 2009). The USA also clarified during WP 1B meetings that they are using the TVWS on non-protection, non-interference basis (Chairman of Working Party 1B, 2013).

It is believed that the timing of the introduction of TVWS was not suitable for a variety of reasons. Several countries had not finalized their switchover plans, providing uncertainty for TVWS operators. Additionally, as WRC-12 allocated the 700 MHz band for mobile services in addition to broadcasting, several countries had to refarm the band in order to introduce mobile service and replan the rest of the UHF band to accommodate TV channels. Another obstacle was the fear that if TVWS devices are widely deployed within a broadcasting environment, this may deter the deployment of mobile service if the allocation changed in future (Freyens & Loney, 2011a, b).

TVWS may have stood a better chance if it had been launched before the RRC-06 conference, when analogue terrestrial broadcasting was well established around the world. However, the consecutive mobile allocations in the 800 MHz and 700 MHz spectrum bands by, respectively, WRC-07 and WRC-12 significantly reduced the chances for TVWS in the whole of the UHF band. Such an allocation would have also threatened broadcasters, so they were, unsurprisingly, quite reluctant to allow any operations in the already reduced UHF band frequencies (El-Moghazi et al., 2016).

El-Moghazi et al. (2016) examined the restrictions placed by the international spectrum management regime on TVWS as an example of opportunistic access. They showed that the international regime does not prevent the adoption of opportunistic access in TVWS. However, there are different aspects of restriction that would influence a regulator's decision in this matter. In essence, these restrictive elements are dependent on a country's relationship with its neighbours. Restrictive elements include not having a service allocation where TVWS devices can operate, having additional mobile allocation and the a priori planning of the broadcasting service in the UHF band. On the other hand, TVWS can operate on a non-interference basis according to the ITU-R Article 4.4, and unlicensed operation is allowed by the RR on the condition of not causing interference.

6.5 Summary

In this chapter, we have highlighted the elements within international spectrum management that are related to spectrum usage rights that are important. These rights may be exclusive, where spectrum users have individual access to the spectrum, or collective, where there is common access to spectrum without exclusive property rights. A third type of rights is easement, where one entity allows other users access to the spectrum that is owns or has licensed as long as they do not cause harmful interference to its services. The two types of spectrum easements accommodate overlay where devices access the spectrum in the geographical, temporal or frequency gaps in the licensed user's transmission. In contrast, underlay is where a secondary user will transmit at low power levels within the noise floor of licensed spectrum.

The concept of spectrum usage rights is neither explicitly mentioned nor defined in the RR. However, the rights of spectrum access are organized through the radiocommunication service allocation framework that prioritizes access to the spectrum among the different services allocated within the band. This includes exclusive primary service, sharing between primary and secondary services, operation on NIB according to RR Article 4.4 or sharing between coprimary services and between cosecondary services in the same band.

There are several examples of spectrum uses that is worth highlighting to explore how they fit within the RR framework. For instance, ISM applications have several spectrum allocations where it is required that radiation from equipment used for ISM applications is minimal and that outside the bands designated for ISM equipment radiation levels are such they do not cause harmful interference to radiocommunication services. In some of these allocated bands, such as the 2400–2500 MHz band, radiocommunication services operating within these bands must accept harmful interference which may be caused by ISM applications.

SRD are not regarded as a radio service and do not have primary or secondary status. Accordingly, they operate, as outlined by RR Article 4.4, on a non-interference non-protection basis. WRC-12 examined the effect of emissions from SRDs on radiocommunication services, and the studies concluded that there was no need to modify the RR. Generally speaking, WLAN devices were designated by regulators on a non-protection, non-interference basis with low powers (Anker & Lemstra, 2011). However, WRC-03 decided to allocate the 5150–5350 and 5470–5725 MHz bands on a primary basis to mobile services for the implementation of wireless access systems including RLANs.

Regarding CRS, WRC-07 issued Resolution 956 that invited the ITU-R to study measures such as the need for a worldwide harmonized CPC or to create a database that can assist in the determination of local spectrum usage. WRC-12 did not decide on any particular measure with regard to CRS, with CRS being recognized as technologies and not radiocommunication services, and that any radio system implementing CRS technology should operate in accordance with the RR, and that the use of CRS does not exempt administrators from their obligations. TVWS refers to those geographical interleaved vacant frequencies in broadcasting spectrum. TVWS can operate on a non-interference basis according to the ITU-R Article 4.4, and unlicensed operation is allowed by the RR on the condition of not causing interference. However, there are international elements that may restrict TVWS operations such as not having a service allocation where TVWS devices can operate, having additional mobile allocation and the a priori planning of the broadcasting service in the UHF band.

References

Al-Rashedi, N. (2014). *Existing SRD related ITU-R deliverables.* ITU Workshop on Short Range Devices and Ultra Wide Band, Geneva.

Almarzooqi, M. S. (2014). *Use of TV white spaces by cognitive radio systems: The UAE views.* ITU-R SG 1/WP 1B Workshop: Spectrum Management issues on the use of White Spaces by Cognitive Radio Systems, Geneva.

Anker, P. (2010). *Cognitive radio, the market and the regulator.* IEEE Dyspan 2010 Conference, Singapore.

Anker, P., & Lemstra, W. (2011). Governance of radio spectrum: License exempt devices. In W. Lemstra, V. Hayes, & J. Groenewegen (Eds.), *The innovation journey of Wi-Fi: The road to global success*. Cambridge University Press.
APT. (2011). *Common Proposals for the Work of the Conference. Agenda Item 1.19.* World Radiocommunication Conference (WRC-12), Geneva.
Ard-Paru, N. (2013). *Implementing Spectrum Commons: Implications for Thailand.* Chalmers University of Technology. Ph.D. Thesis.
ASMG. (2011). *Arab States Common Proposals. Common Proposals for the Work of the Conference. Agenda Item 1.19.* World Radiocommunication Conference (WRC-12), Geneva.
ATU. (2011). *African Common Proposals for the Work of the Conference. Agenda Item 1.19.* World Radiocommunication Conference (WRC-12), Geneva.
Benkler, Y. (2002). Some economics of wireless communications. *Harvard Journal of Law and Technology, 16*(1), 25–83.
Benkler, Y. (2012). Open wireless vs. licensed spectrum: Evidence from market adoption. *Harvard Journal of Law and Technology, 26*(1), 69–163.
Billquist, S. (2008). ITU considers how to stimulate global development of cognitive radio. *PolicyTracker*. Retrieved on April 30, 2012, from www.policytracker.com
Billquist, S. (2010). WRC preparatory talks look positive for cognitive proponents. *PolicyTracker*. Retrieved on April 30, 2012, from www.policytracker.com
Cave, M., & Webb, W. (2003). Designing property rights for the operation of spectrum markets. *Papers in Spectrum Trading*(1).
Cave, M., & Webb, W. (2012). The unfinished history of usage rights for spectrum. *Telecommunications Policy, 36*(4), 293–300.
CEPT. (2000). WRC-2000: European common proposals for the work of the conference. Retrieved from www.itu.int
CEPT. (2011). *European Common Proposals for the Work of the Conference, Part 19, Agenda Item 1.19,* World Radiocommunication Conference (WRC-12), Geneva.
Chaduc, J., & Pogorel, G. (2008). *The radio spectrum.managing a strategic resource.* ISTE Ltd.
Chairman of Working Party 1B. (2013). *Report on the meeting of working party 1B.* Retrieved from www.itu.int
Chairman of Working Party 1B. (2014). *Report on the meeting of working party 1B.* Retrieved from www.itu.int
Chairman of Working Party 1B. (2015). *Report on the meeting of working party 1B.* Retrieved from www.itu.int
CITEL. (2012). *Inter-American Proposals for the Work of the Conference. Agenda Item 1.19.* World Radiocommunication Conference (WRC-12), Geneva.
Deffains, B. (2013). *Spectrum property rights: from theory to policy.* In *17th Annual Conference of The International Society for New Institutional Economics*, Florence.
Demsetz, H. (1967). Toward a theory of property rights. *The American Economic Review, 57*(2), 347–359.
ECC. (2011). *Description of practices relative to trading of spectrum usage rights* (ECC Reports). Retrieved on August 1, 2021, from https://docdb.cept.org
ECC. (2013). *ECC reports 205: Licensed shared access (LSA)* (ECC Reports). Retrieved on August 1, 2021, from https://docdb.cept.org
El-Moghazi, M., Whalley, J., & Irvine, J. (2012). Allocating spectrum: Towards a common future. In *IEEE Symposium on New Frontiers in Dynamic Spectrum Access Networks*, Bellevue.
El-Moghazi, M., Whalley, J., & Irvine, J. (2016). International radio spectrum management regime: Restricting or enabling opportunistic access in the TVWS? ITS 2016 Conference, Cambridge.
ERC. (1992). *Recommendation T/R 22-06: Harmonised radio frequency bands for high performance radio local area networks (HIPERLANs) in the 5 GHz and 17 GHz frequency range.* https://docdb.cept.org
Eurostrategies and LS-Telecom. (2007). *Study on radio interference regulatory models in the European community.* Retrieved on August 1, 2021, from https://op.europa.eu/

Evci, C., & Fino, B. (2005). *Limits and paradoxes in radio spectrum management*. International Union of Radio Science General Assembly, New Delhi, India.

Forge, S., Horvitz, R., & Blackman, C. (2012). *Perspectives on the value of shared spectrum access: Final report for the European commission*. http://ec.europa.eu

Freyens, B. P., & Loney, M. (2011). *Digital switchover and regulatory design for competing white space usage rights*. IEEE Dyspan 2011 Conference, Aachen.

Freyens, B. P., & Loney, M. (2011). Projecting regulatory requirements for TV white space devices. In S. J. Saeed & R. A. Shellhammer (Eds.), *TV white space spectrum technologies, regulations, standards, and applications*. CRC Press.

Hayes, V., & Lemstra, W. (2008). *License-exempt: The emergence of Wi-Fi*. The Genesis of Unlicensed Wireless Policy: How Spread Spectrum Devices Won Access to License-Exempt Bandwidth, Virginia, USA.

Hazlett, T. W. (2006). The spectrum-allocation debate: An analysis. *IEEE Internet Computing, 10*(5), 68–74.

Hazlett, T. W., & Muñoz, R. E. (2006). Spectrum allocation in Latin America: An economic analysis. *George Mason Law & Economics Research Paper, 6*(44), 261–278.

Hohfeld, W. (1913). Some fundamental legal conceptions as applied in judicial reasoning. *Yale Law Journal, 23*(16), 16–59.

Holland, O., Nardis, L. D., Nolan, K., Medeisis, A., Anker, P., Minervini, L. F., Velez, F., Matinmikko, M., & Sydor, J. (2012). *Pluralistic licensing*. IEEE Dyspan 2012 Conference, Bellveu.

Horne, W. D. (2003). *Adaptive spectrum access: Using the full spectrum space*. TPRC, Virginia.

Horvitz, R. (2013). Geo-database management of white space vs. open spectrum. In E. Pietrosemoli & M. Zennaro (Eds.), *Tv white space: A pragmatic approach*. ICTP.

ITU-R. (1932). *General radiocommunication regulations*. International Telecommunication Convention. ITU.

ITU-R. (1938). *General radiocommunication regulations*. International Telecommunication Convention. ITU.

ITU-R. (1979). WRC-79 resolution 63. Relating to the protection of radiocommunication services against interference caused by radiation from industrial, scientific and medical (ISM) equipment. In *Final Acts—World Radiocommunication Conference (WRC-79)*. ITU.

ITU-R. (2003a). ITU-R recommendation M.1450–2. Characteristics of broadband radio local area networks. In *M Series. mobile, radiodetermination, amateur and related satellite services*. ITU.

ITU-R. (2003b). Resolution 229: Use of the bands 5150–5250 MHz, 5250–5350 MHz and 5470–5725 MHz by the mobile service for the implementation of wireless access systems including radio local area networks. In *Provisional Final Acts—World Radiocommunication Conference (WRC-03)*. ITU.

ITU-R. (2007a). Resolution 63: Protection of radiocommunication services against interference caused by radiation from industrial, scientific and medical (ISM) equipment. In *WRC-07 final acts*. ITU.

ITU-R. (2007b). Resolution 805. Agenda for the 2011 world radiocommunication conference. In *Provisional Final Acts—World Radiocommunication Conference (WRC-07)*. ITU.

ITU-R. (2007c). Resolution 953: Protection of radiocommunication services from emissions by short-range radio devices. In *WRC-07 final acts*. ITU.

ITU-R. (2007d). Resolution 956. Regulatory measures and their relevance to enable the introduction of software-defined radio and cognitive radio systems. In *Provisional Final Acts—World Radiocommunication Conference (WRC-07)*. ITU.

ITU-R. (2008a). Appendix 4, Annex 1: Characteristics of stations in the terrestrial services. In *Radio regulations* (Vol. 1). ITU.

ITU-R. (2008b). Article 15: Interferences. In *Radio Regulations*. ITU.

ITU-R. (2009a). ITU-R report SM.2152. Definitions of software defined radio (SDR) and cognitive radio system (CRS). In *SM series. Spectrum management*. ITU.

ITU-R. (2009b). *ITU-R report SM.2152. Definitions of software defined radio (SDR) and cognitive radio system (CRS)*. ITU.
ITU-R. (2011a). CPM report on technical, operational and regulatory/procedural matters to be considered by the 2012. World Radiocommunication Conference. ITU.
ITU-R. (2011b). ITU-R recommendation M.1652-1. Dynamic frequency selection in wireless access systems including radio local area networks for the purpose of protecting the radiodetermination service in the 5 GHz band. In *M series. Mobile, radiodetermination, amateur and related satellite services*. ITU.
ITU-R. (2011c). Report ITU-R M.2225: Introduction to cognitive radio systems in the land mobile service. In *M series. mobile, radiodetermination, Amateur and related satellite services*. ITU.
ITU-R. (2012a). *Frequently asked questions, notification*. Retrieved on April 26, 2012, from http://www.itu.int
ITU-R. (2012b). ITU-R resolution 58: Studies on the implementation and use of cognitive radio systems. In *Resolutions radiocommunication assembly (RA-12)*. ITU.
ITU-R. (2012c). *ITU-R resolution 58: Studies on the implementation and use of cognitive radio systems*. Retrieved from www.itu.int
ITU-R. (2012d). WRC-12 recommendation 76 deployment and use of cognitive radio systems. ITU.
ITU-R. (2013). *Coordination committee for vocabulary (CCV) liaison statement to ITU-R working parties 1B and 6A: Definition and translation of the term "WHITE Space" and related terms*. www.itu.int
ITU-R. (2015). *Provisional Final Acts—World Radiocommunication Conference (WRC-15)*. ITU.
ITU-R. (2018). Recommendation ITU-R SM.1896-1: Frequency ranges for global or regional harmonization of short-range devices. In *ITU-R recommendations: SM-series*. ITU.
ITU-R. (2019a). Administrative circular CA/251: Results of the first session of the Conference Preparatory Meeting for WRC-23 (CPM23-1). Retrieved from www.itu.int
ITU-R. (2019b). Report SM.2153: Technical and operating parameters and spectrum use for short-range radiocommunication devices. In *ITU-R reports: SM-series*. ITU.
ITU-R. (2020a). Appendix 18: (REV.WRC-19) Table of transmitting frequencies in the VHF maritime mobile band. In *Radio regulations*. ITU.
ITU-R. (2020b). Appendix 27: Frequency allotment plan for the aeronautical mobile (R) service and related information. In *Radio Regulations*. ITU.
ITU-R. (2020c). Article 11: Notification and recording of frequency assignments. In *Radio Regulations*. ITU.
ITU-R. (2020d). Article 13: Instructions to the Bureau. In *Radio Regulations*. ITU.
ITU-R. (2020e). Article 15: Interferences. In *Radio Regulations*. ITU.
ITU-R. (2020f). Article 21: Terrestrial and space services sharing frequency bands above 1 GHz. In *Radio Regulations*. ITU.
ITU-R. (2020g). Article 52: Special rules relating to the use of frequencies. In *Radio Regulations*. ITU.
ITU-R. (2020h). Article 5: Frequency allocations. In *Radio Regulations*. ITU.
ITU-R. (2020i). Resolution 750: (Rev.WRC-19) Compatibility between the Earth exploration—Satellite service (passive) and relevant active services In *Radio regulations*. ITU.
Lehr, W., & Crowcroft, J. (2005). Managing shared access to a spectrum common. In *IEEE Symposium on New Frontiers in Dynamic Spectrum Access Networks*, Baltimore.
Lemstra, W., Groenewegen, J., & Hayes, V. (2011). The case and the theoretical framework. In W. Lemstra, V. Hayes, & J. Groenewegen (Eds.), *The innovation journey of Wi-Fi: The road to global success*. Cambridge University Press.
Lemstra, W., Links, C., Hills, A., Hayes, V., Stanley, D., Heijl, A., & Tuch, B. (2011). Crossing the chasm: The Apple airPort. In W. Lemstra, V. Hayes, & J. Groenewegen (Eds.), *The innovation journey of Wi-Fi: The road to global success*. Cambridge University Press.
Negus, K. J., & Petrick, A. (2009). History of wireless local area networks (WLANs) in the unlicensed bands. *Info, 11*(5), 36–56.

Newlands, M. (2009). CEPT 'white space' report leaves cognitive radio issues unresolved. *PolicyTracker*. Retrieved on March 19, 2014, from https://www.policytracker.com

NTRA of Egypt. (2014). *Views on TV white spaces (TVWS)* 2nd AfriSWoG Meeting, Nairobi.

Ofcom. (2006). *Spectrum usage rights: Technology and usage neutral access to the radio spectrum*. Retrieved from https://www.ofcom.org.uk/

Ofcom. (2008). *Spectrum usage rights: A guide describing SURs*. http://stakeholders.ofcom.org.uk

Ofcom. (2012). *UK Report of the ITU World Radio Conference (WRC) 2012*. Retrieved from http://stakeholders.ofcom.org.uk

QinetiQ. (2005). *Bandsharing concepts*. Retrieved from www.spectrumaudit.org.uk

RCC. (2011). *Common Proposals by the RCC Administartions on WRC-12 Agenda Item 1.19*, World Radiocommunication Conference (WRC-12), Geneva.

RSPG. (2011). *Report on Collective Use of Spectrum (CUS) and Other Spectrum Sharing Approaches*. Retrieved from http://rspg.groups.eu.int

Savage, J. (1989). *The politics of international telecommunications regulation*. Westview Press.

Shannon, C. E. (1948). A mathematical theory of communication. *The Bell System Technical Journal, 27*, 379–423.

Vany, A. S. D., Eckert, R. D., Meyers, C. J., O'Hara, D. J., & Scott, R. C. (1969). A property system for market allocation of the electromagnetic spectrum: A legal-economic-engineering study. *Stanford Law Review, 21*(6), 1499–1561.

Vries, P. D., & Sieh, K. A. (2012). Reception-oriented radio rights: increasing the value of wireless by explicitly defining and delegating radio operating rights. *Telecommunications Policy, 36*(7), 522–530.

Weiser, P., & Hatfield, D. (2008). Spectrum policy reform and the next frontier of property rights. *George Mason Law Review, 60*(3), 549–609.

Werbach, K. (2004). Supercommons: Toward a unified theory of wireless communication. *Texas Law Review, 82*, 863–973.

Whittaker, M. (2007). *Commercial Certainty in Spectrum Right Formulation*. Retrieved from www.futurepace.com.au

Yahoo Finance. (2016). First part of the FCC incentive auction sets hefty price. In *Yahoo finance*. Retrieved on October 8, 2016, from http://finance.yahoo.com

Zhao, Q., Davis, C., & Sadler, B. M. (2007). A survey of dynamic spectrum access. *IEEE Signal Processing Magazine, 24*(3), 79–89.

Chapter 7
Spectrum Rights Assignment

No transmitting station may be established or operated by a private person or by any enterprise without a license issued in an appropriate form and in conformity with the provisions of these Regulations by or on behalf of the government of the country to which the station in question is subject.

Article 18 of the RR (2020)

7.1 Introduction

Frequency assignment is considered as the last stage of spectrum management where the regulator decides on how to award the frequencies to the operators or end-users. On first glance, it may appear that the ITU-R should not be involved in national matters such as assignment frequencies as it is mostly concerned with international coordination and cross-border issues. However, this is not totally accurate. In fact, since the early days of the RR, there have been several articles that deal with spectrum assignments nationally in addition to several ITU-R recommendations and reports that focus on issues such as spectrum pricing and the refarming of frequencies. For instance, Article 18 of the ITU-R RR, as quoted above, required the licensing of stations that are in conformity with the RRs (ITU-R, 2020d). In addition, ITU-R reports such as Report SM.2012 discuss non-market-based methods (e.g. comparative processes and lotteries) and market-based methods (e.g. auction and spectrum trading) that could be used to assign spectrum (ITU-R, 2018a).

With this in mind, this chapter addresses spectrum rights assignment and examines the extent to which it is related to the international spectrum management framework. The chapter begins with a discussion of the different models of spectrum assignment before exploring the different ways that the RR and other ITU-R regulations may influence the process of spectrum assignment nationally. Section 7.4 focuses on the process of notifying spectrum assignments to the ITU and the importance of this for national administrations. The next four sections then address the different components of the assignment process—licensing, pricing, assignment conditions

M. A. El-Moghazi and J. Whalley, *The International Radio Regulations*,
https://doi.org/10.1007/978-3-030-88571-7_7

and spectrum trading—and how these are influenced by ITU regulations. Finally, the chapter elaborates the latest innovative spectrum assignments measures.

7.2 Radio Spectrum Assignment Models

Assignment could be described, according to one ITU report, as '*the regulator's national exercise following allocating the band to a particular service in the RR*' (ITU-R, 2018a). In other words, it is acknowledged that it is a step that follows after agreeing on a specific radiocommunication service allocation. It is also recognized that it is not an international matter to be decided within the ITU (ITU-R, 2018a). Instead, it is a national matter that is usually undertaken by national regulators. However, the assignment process is a complicated process as we shall explain in this chapter.

Firstly, the assignment process accommodates three components: authorization, assignment mechanisms and procedures and usage conditions. The authorization stage can be individual or general, and it is often related to how the spectrum is licensed. More specifically, in the general authorization model, there is usually no form of protection against harmful interference from an authorized use of spectrum. This model is more suitable when there is no need to coordinate between large number of users in the same area, and no harmful interference is expected with neighbouring countries. In the individual authorization model, the license entails protection against harmful interference (RSPG, 2009).

The assignment mechanism could take one of several forms such as the administrative assignment method where the regulator decides how the spectrum will be used by designating appropriate uses, technologies and users (OECD, 2006). The regulator also determines how long the license will run for and its associated obligations (WIK, 2006). There are two types of awarding the license under this model: first come first served (FCFS) and beauty contest. Using the FCFS method, license applications are dealt with in the order of their receipt, and the license is granted when the applicant fulfils the application criteria (ERC, 1998). It is generally used when there is no shortage of spectrum, and it has the advantage of low administrative costs and producing quick assignments (RSPG, 2009). In contrast, a beauty contest requires the regulator to choose the winning applicant via a competitive process using comparative criteria previously decided (Hatfield, 2005). This method is widely used to achieve non-market public interest benefits and to address policy objectives that the market fails to take into account (FCC, 2002).

If demand for the spectrum exceeds supply, the regulator could choose from the applicants using specific criteria to compare and differentiate between them. Such criteria could be comparative hearings, lotteries or auction. In some cases, a combination of these criteria is used, for example, initially using comparative hearings to initially screen the applicants then applying an auction (ITU-R, 2018a). Auctions are the application of the market in the assignment process and could be described as a contest for a license with a focus on the price bid. There have been several developments in the auction process so different elements can be included aside from the price of the license such as deployment roadmaps or coverage obligations.

The use of auctions was first suggested in 1959 by Ronald Coase who argued that auctioning spectrum to the highest bidder was the most effective method of assignment (Economist, 2004). Auctions are usually associated with the use of spectrum trading in a secondary market (Baumol & Robyn, 2006) and are held to create incentives for spectrum users to apply their spectrum to the highest-valued uses as determined by the market (OECD, 2006). Secondary markets or spectrum trading allows spectrum users to trade their licenses if there is a new user who places more value on the spectrum so that a more economically efficient use of the spectrum is achieved (Cave et al., 2007).

General authorization entails exempting from licensing. License exempt is often referred to as an open or free approach (Lehr, 2005). In such an approach, radio interference can be considered as a technological problem. Given the availability of smart radio and antennas, interference is resolved automatically by the users themselves with no intervention by the regulator (Economist, 2004). The license exempt approach is generally suitable for frequency bands where scarcity is relatively low, and the transaction costs associated with market-based negotiation of access rights are relatively high (FCC, 2002). License conditions could include service requirements, coverage requirements and obligation requirements (RSPG, 2009).

Combining these aforementioned components results in different types of spectrum access such as Dynamic Spectrum Access (DSA), Licensed Shared Access (LSA), private commons, license exempt or exclusive individual licensing (LS telcom AG et al., 2017).

The assignment of spectrum is often supplemented by imposing a charge for spectrum use. Charging facilitates the allocation of scarce resources and conveys information about the supply and demand for spectrum (Mueller, 1982). Moreover, charging is a major factor in utilizing the spectrum as too high a price might lead to underutilization of spectrum, while too low a price might lead to hoarding and congestion (Cave et al., 2007). Ideally, charging should be considered as a tool for rationalizing the usage of the spectrum and not as a way to generate revenue for a government (Youssef et al., 1995).

The type of spectrum pricing is related directly by the way spectrum is assigned and whether it is via administrative measures, auction or traded in the market. Administrative pricing is associated with the 'command and control' spectrum management approach, and there are different methods in use for such pricing. These include differential, incentive, opportunity costs, periodic administrative cost recovery, shadow, spectrum refarming and user profit pricing (ITU-R, 2018a).

One of the most widely adopted administrative pricing techniques is the incentive type based on opportunity cost that is known as 'Administrative Incentive Pricing' (AIP). Licenses are issued through an administrative process, while at the same time, the fees are based on the opportunity cost. The method provides an incentive to the licensee to return excess spectrum or to use spectrum more efficiently. AIP is intended to ensure that decisions by spectrum planners and users reflect the value of spectrum, not just to themselves but also to other users (DTI, 2007). Pricing spectrum at a value closer to its economic value provides a disincentive to hoarding spectrum (Smith-NERA, 1996). Finally, the reassignment of spectrum is a process needed for refarming the spectrum where the regulator recovers the spectrum from its existing users and assigns it to new uses (ERC, 1998). Spectrum refarming could be achieved through different measure such as license termination (El-Moghazi et al., 2008).

Although in this section we have provided an overview on the different assignment models and stages, it is still unclear whether ITU regulations in general or the RR in particular have an influence on the spectrum assignment given that it is usually considered to be a national task and that the RR do not deal with the process of authorizing frequency usage (Anker & Lemstra, 2011). In the following sections, we shall explore the pillars of spectrum assignments that may be influenced by the different elements of the international spectrum management including the RR.

7.3 The Influence of the RR on Spectrum Assignments

The RR accommodate a definition of the process of assignment in Article 1.18 as '*[a]uthorization given by an administration for a radio station to use a radio frequency or radio frequency channel under specified conditions*' (ITU-R, 2020a). Another important provision in the RR is Article 4.2, which states that these assignments which could cause harmful interference should be in accordance with the RR. An earlier version of this article, following the Washington Conference of 1927, has different wording where assignments are under the sovereign authority of governments on the condition of not causing interference to other countries. It does not explicitly mention that these assignments should be in accordance with the RR.

Other provisions related to the technical performance of frequency assignments nationally are within Article 4 of the RR. Most importantly, Article 4.1 urges Member States to '*limit the number of frequencies and the spectrum used to the minimum essential to provide in a satisfactory manner the necessary services. To that end they shall endeavor to apply the latest technical advances as soon as possible*'. Another important provision is Article 4.11, which is related to the usage of frequencies in the HF band, which are capable of covering long distances, urges Member States to utilize these frequencies in the case of need for long distance communications. In other words, such provisions encourage maximizing spectrum utilization efficiency and to follow the latest advancement in wireless communications. The challenge, of course, in such cases is that there are no defined measures to assess the applicability of such provisions.

Arguably one of the most important provisions in the RR is Article 4.4. This article states that '*[a]dministrations of the Member States shall not assign to a station any frequency in derogation of either the Table of Frequency Allocations in this Chapter or the other provisions of these Regulations, except on the express condition that such a station, when using such a frequency assignment, shall not cause harmful interference to, and shall not claim protection from harmful interference caused by, a station operating in accordance with the provisions of the Constitution, the Convention and these Regulations*' (ITU-R, 2020b).

In other words, such provision allows operating in a way not aligned with the RR on the condition of not causing harmful interference to those stations operating according to the RR or claiming protection from harmful interference from these stations. This, in theory, provides the ultimate flexibility and freedom to act independently of the RR considering the previously mentioned conditions. In fact, in several cases during WRCs where a country or a group of countries failed to acquire specific service allocation through footnote, they are advised to operate the service according to the RR Article 4.4. However, this is usually rejected as such an operation comes with an expensive price in terms of certainty, and it is considered as the last resort for wireless operations. This is because it is rare to find a manufacture or operator who is willing to invest in providing wireless services without any protection against interference and without the right to even cause any claimed interference to other services.

It is important to highlight which ITU study groups address assignment procedures in a direct way even if their outputs are not reflected in the RR. One group is ITU-R Working Party (WP) 1B, which is responsible of spectrum management methodologies and economic strategies which are contrast to most other ITU groups that are technical in nature (ITU-R, 2020f). The current scope of WP 1B includes long-term strategies for spectrum utilization, alternative methods of national spectrum management, spectrum redeployment as a method of national spectrum management and economics aspects on spectrum management (ITU-R, 2015b). Another important ITU-R document is the Handbook on National Spectrum Management which covers the main elements of spectrum management such as planning and authorization (ITU-R, 2015a).

Another important non-mandatory document that addresses many issues related to the assignment of frequencies nationally is ITU-R Report SM.2012, which was first released in 1998. The report demonstrates different approaches to national spectrum management with a focus on pricing and refarming (ITU-R, 2018a).

Not all activities related to spectrum assignment are conducted solely by the ITU-R. For instance, in response to Resolution 9 of the ITU World Telecommunication Development Conference (WTDC), a joint ITU-R/ITU-D Group was established with a focus on the participation of countries, particularly developing countries, in spectrum management. The resolution is usually revised at each WRDC conference, and its latest version following the WTDC of 2017 in Argentina focused on issues such as innovative ways of spectrum licensing (e.g. light licensing and authorized shared access/licensed shared access) (ITU-D, 2017).

7.4 Notifications of Spectrum Assignments

The end of the spectrum assignment process is the beginning of another process that directly involves the ITU-R that is called frequency notification. Following the frequency assignments nationally, administrations can notify these assignments for recording in the Master International Frequency Register (MIFR). Article 4 clarifies those cases where notification of frequency assignments is required, including those which are capable of causing harmful interference to any service of another administration, or in cases where the assignment is to be used for international radiocommunication. According to RR Article 8.3, Member States should take into account these recorded assignments when they are making their own assignments in order to avoid harmful interference.

Notification is important for various reasons. Firstly, it secures international recognition for these assignments, as other countries take these assignments into consideration when assigning their own frequencies in order to avoid harmful interference to those assignments recorded in the MIFR. Secondly, it reflects the actual use of some frequency bands which could be important for discussions at WRCs. More specifically, if there is an agenda item that addresses a new service allocation in a particular frequency band, the assignments registered in the MIFR would reflect the actual usage of the band by the incumbent or existing service. If there are few assignments this could be an indication that sharing, and thus the introduction of a new service, would be relatively easy. For instance, one of the reasons for not studying the 28 GHz by WRC-19 is the large number of satellite network filings that have been made and the number of satellites already in operation (Roberti, 2019).

There are specific categories of frequencies that can be notified to the ITU-R BR. This includes those that are used for international radiocommunication and capable of causing harmful interference to any service of another administration. There are cases where it is not possible to notify particular types of assignment (ITU-R, 2020c). This includes assignments to stations in the amateur service or to assignments involving specific frequencies which are prescribed by the RR for common use by terrestrial stations of a given service (ITU-R, 2020e).

7.5 Spectrum Licensing

We previously mentioned that there are two types of spectrum authorization, general and individual, where in the former a license is not required or exempted. Surprisingly, one whole article of the RR focuses on licensing and even instructs when it should be applied. More specifically, Article 18.1 states that '*[n]o transmitting station may be established or operated by a private person or by any enterprise without a license issued in an appropriate form and in conformity with the provisions of these Regulations by or on behalf of the government of the country to which the station in question is subject*' (ITU-R, 2020d). Article 18 is described in one of the ITU-R reports as the link between international and national spectrum management as it ensures the commitment of the national stakeholders to the international treaty (i.e. the RR) that is based on not causing harmful interference (ITU-R, 2018b).

While one may understand why a license is required in cases of international air flights or marine voyages, why is there a need to mention licensing in general on the national level in an international agreement? In order to understand this, perhaps it is useful to trace the origin of licensing in the RR. The analysis by Ard-Paru (2013) shows that the notion of licensing has been around since the first version of the RR in 1906, when it was stated that stations on ships board should have a license issued by the government to which the ship belongs.

The 1927 Washington conference modified the article to include any radioelectric sending station, and also, it included individual persons. The RR of 1982 was slightly modified to indicate that such a station should be in conformity with the RR. The bottom line is that the current article addressing licenses originated almost a century ago. Since then, the concept of licensing on the national level has developed to include different types of licenses (e.g. light licensing, general authorization) without major implications for the RR.

However, it seems that the mention of licensing in the RR could have implications for national licensing. To understand these implications, it is necessary to have a closer look on an interesting survey that was conducted in 2006 to examine the license conditions of Wi-Fi. It was found that Wi-Fi was license exempt in only 67 of the 167 countries that completed the survey. Moreover, a survey by the ITU on international policies on license exempt shows that almost two thirds of the 75 responding countries do not enjoy full license exemptions (Best, 2006). Whether such a result was related to conducting the survey during the early days of Wi-Fi is unclear, but Horvitz (2007) asks an interesting question related to the licensing on the national level and whether that is influenced by the RR provisions. Horvitz (2007) argues that countries where laws prohibit the unlicensed use of the spectrum are largely influenced by the RR Article 18.1 that mandates licensing.

Another example of the influence of the RR on licensing nationally is the discussion on managing the unauthorized operation of earth station terminals deployed within national territories. This began at the Radio Assembly of 2015 when Resolution 64 addressed the increasing demand for global broadband communication via systems such as high-density applications in the fixed-satellite service (HDFSS). HDFSS systems usually are deployed in rapid and ubiquitous manner using small antennas (ITU-R, 2015c). The resolution recognized Article 18 of the RR that states no transmitting station may be established or operated by a private person or by any enterprise without a license in conformity with the RR.

Finally, the resolution invited the relevant ITU-R study groups to conduct studies to examine whether there is a need for possible additional measures to limit uplink transmissions of terminals to those terminals authorized in accordance with the licensing articles in the RR. The issue is simply that in the era of open skies, developed countries have the power to launch satellites that provide broadband services which could be utilized by individual users via relatively small earth station terminals. For developing countries, it is usually difficult for them to monitor the activities of these terminals, especially the uplink activities. Ideally, these stations should acquire licenses from the national regulator, and the satellites should also have permission to cover the territories of these countries. In addition, these terminals could be notified to the ITU-R for purposes of coordination with other services or neighbouring countries.

The next step was to include the study of such an issue under the urgent studies in preparation for WRC-19 under Agenda Item 9.1.7 as stated in Resolution 958 (ITU-R, 2015d). Two issues were examined under the A.I., whether there is a need for possible additional measures in order to limit uplink transmissions of terminals to those authorized terminals, and the possible methods that will assist administrations in managing the unauthorized operation of earth station terminals deployed within its territory.

Studies prior to WRC-19 showed the concern of some countries that there is no clear framework in the RR for administrations to apply their complaints regarding licensing in practice and that there is no clear provision in the RR to address unauthorized transmission of earth stations operating within a given satellite network. On the other hand, there was a view that the RR, in Article 18, contains a clear and unambiguous requirement to operate an earth station only if authorized. The studies have also showed that Article 18.1 of the RR could be implemented by the administrations in different ways of licensing (individual licensing, simplified licensing and voluntary registration of earth stations) and that violations of a national licensing regime could be considered as not being in alignment with the RR. Ultimately, WRC-19 decided in Resolution 22 that the operation of transmitting earth stations within the territory of an country should be conducted only if authorized by them (ITU-R, 2019).

A closer look at the resolution reveals that it addresses two interrelated issues. The first is the authorization of these stations by a national government or regulator. This could be perceived to be a national matter. The second is that in case these stations are not authorized, the satellite operator should not allow these stations to be operated. Such an issue could be considered to be international as it is related to satellite operations that cover the territories of several countries.

Several other observations can also be deducted from the resolution. The first is that there are no clear enforcement measures inherited from the RR for usages that are in compliance with the regulations. That is why the resolution used the vague phrase 'to the extent practical'. However, the resolution also included some procedures to clarify the operation of such a resolution. These include that the country where there is an unauthorized earth station may report such an operation to the country to which the satellite network belongs to. The ITU-R could also be involved to resolve any issues, but again, there is no strict enforcement mechanisms.

7.6 Spectrum Pricing

In the previous section, we highlighted the main ITU-R reports that address national practices of spectrum pricing. Here, we aim to examine whether the RR provisions could influence spectrum prices.

Firstly, the funding of national spectrum management, which is usually contributed to by revenue coming from spectrum pricing, should cover country's international obligations such as participating in national forums (e.g. ITU-R WPs) (ITU-R, 2018a). For instance, an incentive pricing model was developed by ITU experts for the south-east Asia region to calculate spectrum fees on the basis of tangible criteria. The model suggested that the fees should cover the expenditures of the regulator, some of which are related to the participations in international spectrum management activities (e.g. WRCs) (Pavliouk, 2000). Therefore, services which require extensive international coordination may also imply higher administrative fees to cover the regulator's expenses.

Another observation is that even the detailed calculation of the pricing system nationally could be influenced by different ITU-R documents. For instance, the coverage of transmitters is dependent on the ITU-R recommendations of propagation models. Another potential element is the areas occupied of satellite earth station which are dependent on coordination distances determined by the ITU-R. One may even argue that the RR have an indirect influence on spectrum fees where services with more stringent conditions on them may decrease its value and, therefore, the license fees to the operators or the end-users. One example of this is related to the influence of the conditions of EESS protection at the lower band of the 24 GHz where, in the USA, the lower band was auctioned for lower prices than the upper band (Youell, 2019).

Another potential impact is on the scarcity of spectrum for cellular mobile, which has, in turn, an influence on spectrum prices. In particular, the examination of the behaviour of mobile operators during the 3G auctions in Germany and the UK showed that as operators feared being excluded from the market if they were not awarded a license, they behaved irrationally and overbid for the license (French, 2009). Those auctions that focused on maximizing revenue resulted in higher consumer prices and lower service quality (GSMA & LS telcom AG et al., 2017; GSMA & NERA, 2017).

The scarcity of spectrum which lead to the usage of auction is also related to the concept of IMT identification, which has previously been discussed in this book, which limits the use of cellular mobile to mobile service allocated frequencies that are identified as IMT. At the time of the 3G and 4G auctions, there were relatively small amounts of IMT identified frequencies available, and many regulators did not adopt technology neutral policies. Accordingly, operators were under the threat that if they do not obtain frequencies that were identified for IMT or IMT-2000 at that time (e.g. 2000 MHz band), they would be excluded from providing 3G services. However, the situation has significantly changed following WRC-19 that identified more than 15 GHz for IMT. This should contribute to a decrease in spectrum prices regardless of the assignment process. This is supported by the relatively low prices paid for 5G licenses in Europe which are correlated to the amount of spectrum made available (Cave, 2019).

7.7 Assignment Conditions

Whether the assignment of spectrum is licensed or exempt, its usage has some conditions that could be incorporated into the user's license or device configuration. Generally speaking, the detailed technical characteristics of the stations to which frequencies are assigned should adhere to the provisions of the RR. This includes, but is not limited to, unwanted emissions, transmitted power and signal-processing methods as mentioned in Article 3 of the RR.

A study conducted on 5G usage in the millimetre bands in the European Union discussed issues that influence the usage conditions related to the protection of EESS in the 26 GHz band. The main insight of the study is that the restrictive conditions associated with the protection of the passive EESS below 24 GHz could render the lowest part of the 26 GHz band (1.25 GHz) inoperable for 5G, at least outdoors (IDATE & Plum, 2019). These conditions were later discussed at WRC-19 and incorporated into the RR. Accordingly, it is expected that 5G licenses around the world will adhere to these conditions.

Another example of the involvement of the RR in usage conditions nationally is Resolution 229. While the 5150–5350 MHz band was allocated to mobile service on a primary basis by WRC-03, Resolution 229, which was modified by WRC-19, restricts the use of mobile service in the 5150–5250 MHz band to indoor use with limited transmitted power (ITU-R, 2003). In addition, outdoor usage is permitted under restrictive conditions. The resolution recognized that there are measures to achieve these restrictions including the authorization approach and limited application. The resolution requires administrators to take all appropriate measures to control the number of these higher power outdoor WAS/RLANs stations.

Licenses often accommodate conditions related to the refarming of spectrum and change in its use. One of the main reasons for refarming spectrum is that the international harmonization of spectrum may necessitate the reallocation of frequencies. Furthermore, the ITU may decide to allocate a currently occupied frequency band to a different service on a regional or global basis (ERC, 1998).

7.8 Spectrum Trading

As spectrum trading has been an important issue in the literature (Cave & Webb, 2003), we assess in this section whether the RR has an influence on the national application of such processes. In general, the licensing articles in the RR, mainly Article 18.1, require issuing a license and do not forbid the transfer of such a license. But what about services of international nature such as satellites?

One report on the implications of international regulations on spectrum management shows that trading is possible, to introduce flexibility in both service and technology usages, albeit with some restrictions on small geographically countries with high population density and several neighbouring countries (Indepen, 2001). The report views the RR as an international framework that provides a considerable degree of flexibility for national regulator regarding spectrum usage nationally.

It seems, therefore, that flexibility in license conditions is also related and connected to radiocommunication service allocation and that spectrum trading or secondary market of licenses is not limited to changes of the spectrum property owner. In other words, as long as the licensee adheres to the conditions of the ITU-R RR, there is no restriction nationally. However, this does not apply to satellite services. Several provisions of the RR allow for a country to act on behalf of a group of named countries for the purpose of notifying ITU-R of frequency assignments to satellite systems (ITU-R, 2017). In such cases, the country acting on behalf of the group is designated as the notifying country. The satellite system cannot be transferred to another notifying country. In other words, the notified satellite network cannot be traded.

Having said that, the leasing of satellite capacity or unused satellites slots for a limited time period is not prohibited by the ITU-R RR, and there have been proposals for developing countries to lease out or trade their orbit frequency assignments (Levin, 1988). In addition, there have been some suggestions to introduce a secondary market for satellite systems where there are slots reserved for future use in the BSS and FSS frequencies for all ITU countries. In such cases, leasing is possible in the planning of slots where spectrum rights are well-defined (Nozdrin, 2008).

7.9 Innovation in Spectrum Assignments

In this section, we explore the different innovative assignment measures that are suggested in the literature, which is then followed by a discussion of the influence of the RR on these approaches.

There has recently been a trend in reviewing traditional assignment measures adopted in the 3G and 4G eras that focused on providing more incentives for operators to invest and incorporating coverage obligations in the pricing objectives (Cave, 2019; Pogorel & Bohlin, 2017). The assignment process is expected to be different for the case of 5G due to new techniques such as network slicing, which enables the creation of a virtual network in order to provide specific service to different users with particular needs (e.g. hospital, school). Accordingly, there are different approaches of spectrum licensing for what are considered as vertical service network providers (VNSP). This includes allowing the mobile network operators to lease or trade the spectrum to the VNSP. Other options are for the regulator to issue a local license to the VNSP or to allow secondary access in bands already licensed to the incumbent. The fourth option is similar to the traditional virtual mobile operator where the mobile operators provide a dedicated virtual network for the VNSP. The fifth option is to allow unlicensed operation (Vuojala et al., 2020).

Other options for spectrum assignments accommodate a solution for non-exclusive access to the spectrum which was recommended within the EU through general authorization for 5G (LS telcom AG et al., 2017). Ofcom, in 2011, planned to reserve part of the 2.6 GHz band for low-power operations on a shared basis. The authorization process included having the opportunity of bidding for low-power or high-power spectrum lots. Ultimately, no low-power licenses were awarded in the auction. In the USA, in contrast, a three-tiered spectrum sharing system was established in the 3.5 GHz as known as the Citizens Broadband Radio Service (CBRS). The first tier allows exclusive access to the incumbents in the band, while the second tier allows for new actors known as Priority Access Licensees (PALs). The third tier allows accessing the spectrum on an opportunistic access basis conforming to the General Authorised Access (GAA) framework. Coordination is provided through what is called Spectrum Access System (SAS), which protects higher tier users from interference caused by lower tier users. The assignment mechanisms for PAL applicants are through competitive bidding (Beltrán & Massaro, 2018).

Other revolutionary mechanisms for spectrum assignments tend to reconsider the traditional idea of dividing the spectrum into bands of frequencies that may not be related to the nature of spectrum (Werbach, 2004). For example, concepts such as 'spread-spectrum' enable multiple users to access the spectrum. This is based on the theory that a signal can either be sent across a narrow channel at high power or spread across a wide channel at lower power (Shannon, 1948). Moreover, the regulator should focus on controlling the traffic among the equipment (Noam, 1995) and regulating the use of the equipment themselves (Benkler, 1998).

Another mechanism is to adopt a decentralized architecture in spectrum allocation similar to the Internet (Benkler, 1998). An example of this mechanism called spectrum networking database (SND), which facilitates the management of dynamic access to spectrum in a way similar to the domain name system (DNS) used in the management of the Internet (Werbach, 2010). SND is considered as an extension to the white spaces database proposed by the FCC (FCC, 2008) to all spectrum. In addition, there is a suggestion of applying blockchain to radio spectrum management and enable more dynamic spectrum sharing (Weiss et al., 2019). Finally, there have been calls for dedicated unlicensed spectrum where there is no central-planner to assign the spectrum (Lehr, 2004; Lehr & Crowcroft, 2005).

7.10 Summary

In this chapter, we have explored the different ways to assign spectrum and the influence of the RR on these processes. Frequency assignment is considered as the last stage of spectrum management, where the regulator decides on how to aware the frequencies to operators or end-users. There are, however, many different components to how this occurs—there are, for example, three components to the assignment process (authorization, assignment mechanism and usage conditions) as well as two different ways of authorization (individual and general). There are also two different ways of assigning spectrum under the administrative assignment process (FCFS and beauty contest). What this entails in practice is that countries can vary quite significantly in how they award spectrum.

There are several provisions within the RR related to the assignment process that highlight the importance of limiting the number of frequencies and the spectrum used to the minimum necessary to provide the services in a satisfactory fashion. It is, however, worth remembering that Article 4.4 does allow a country to deviate away from the RR as long as this does not cause harmful interference to those operating in accordance with the RR or results in them seeking protection from the harmful interference caused by these stations. This should come as no surprise, as one of the key overarching guiding principles of the RR is that no harmful interference should occur.

This chapter has also demonstrated how the RR provisions could influence, albeit indirectly, spectrum prices though adopting the pricing models proposed by the ITU-R or by the usage conditions contained within the RR. Another potential impact is through the scarcity of the spectrum for cellular mobile which has, in turn, an influence on spectrum prices and is related to IMT identification. Regardless of whether the assigned spectrum is licensed or exempt, its usage may contain conditions that could be incorporated into the user's license or device configuration. These may be related to provisions within the RR as illustrated by, for example, the cases of IMT in the 26 GHz and RLAN in the 5 GHz bands.

There are certainly ways of improving the contribution of the ITU to the assignment activities that occur nationally, though the assignment process remains a national matter. Many countries, especially developing ones, unfortunately, lack some of the competences needed to manage spectrum auctions. The World Radiocommunication Seminars (WRS), which are held on a biennial basis by the ITU-R, offer a place where the ITU can provide the necessary support to assist its Member States to assign spectrum and accommodate the usage conditions contained within the RR. This would also assist policymakers at the ITU understand how their decisions internationally, in forums such as WRC, have an influence on national developments.

References

Anker, P., & Lemstra, W. (2011). Governance of radio spectrum: License exempt devices. In W. Lemstra, V. Hayes, & J. Groenewegen (Eds.), *The innovation journey of Wi-Fi: The road to global success*. Cambridge University Press.

Ard-Paru, N. (2013). *Implementing spectrum commons: Implications for Thailand*. Ph.D. Thesis, Chalmers University of Technology. Ph.D. Thesis.

Baumol, W., & Robyn, D. (2006). *Toward an evolutionary regime for spectrum governance: Licensing or unrestricted entry? In*. AEI Brookings Joint Center for Regulatory Studies.

Beltrán, F., & Massaro, M. (2018). Spectrum management for 5G: Assignment methods for spectrum sharing. In *29th European Regional ITS Conference*, Trento.

Benkler, Y. (1998). Overcoming agoraphobia: Building the commons of the digitally networked environment. *Harvard Journal of Law and Technology, 11*(2), 1–113.

Best, M. (2006). *A global survey of spectrum license exemptions.* Telecommunications Policy Research Conference.

Cave, M. & Webb, W. (2003). Designing property rights for the operation of spectrum markets. *Papers in Spectrum Trading.* no. 2, University of Warwick (UK), Warwick Business School.

Cave, M. (2019). *Optimizing spectrum assignments to deliver expansive 5G connectivity.* Retrieved from https://www.ericsson.com

Cave, M., Doyle, C., & Webb, W. (2007). *Essentials of modern spectrum management.* Cambridge University Press.

DTI. (2007). *Forward look—a strategy for management of major public sector spectrum holdings, UK spectrum strategy committee in consultation with OFCOM.* Retrieved from www.dti.gov.uk

Economist, T. (2004). On the same wavelength: Special report spectrum policy. *The Economist, 372*(8388), 61–63.

El-Moghazi, M., Whalley, J., & Curwen, P. (2008). Is Re-farming the Answer to the Spectrum Shortage Conundrum? *Management Science Working Paper.* University of Strathclyde.

ERC. (1998). *ERC report 53: Report on the introduction of economic criteria in spectrum management and the principles of fees and charging in the CEPT.*

FCC. (2002). *Report of the spectrum policy task force.* Retrieved from www.fcc.gov

FCC. (2008). *Unlicensed operation in the TV broadcast bands, second report and order and memorandum opinion and order, ET docket no. 04–186, ET Docket No. 02–380, FCC 08–2360.* Retrieved from www.fcc.gov

French, R. D. (2009). Governance and game theory: When do franchise auctions induce firms to overbid? *Telecommunications Policy, 33*(3–4), 164–175.

GSMA & NERA. (2017). *Effective spectrum pricing: Supporting better quality and more affordable mobile services.* Retrieved from https://www.gsma.com

Hatfield, D. (2005). Spectrum management reform and the notion of the spectrum commons. *Southern African Journal of Information and Communication, 4*, 1–12.

Horvitz, R. (2007). Beyond LICENSED VS. unlicensed: Spectrum access rights continua. In *ITU Workshop on Market Mechanisms for Spectrum Management*, Geneva.

IDate & Plum. (2019). *Study on using millimetre waves bands for the deployment of the 5G ecosystem in the Union.* Retrieved August 1, 2021 from https://op.europa.eu/

Indepen, A. (2001). *Implications of international regulation and technical considerations on market mechanisms in spectrum management: Report to the independent spectrum review.* Retrieved from http://www.ofcom.org.uk

ITU-D. (2017). Resolution 9 (Rev. Buenos Aires, 2017): Participation of countries, particularly developing countries, in spectrum management. In *WDTC-2017 Final Acts.* ITU.

ITU-R. (2020f). *Working party 1B (WP 1B)—Spectrum management methodologies and economic strategies.* Retrieved September 20, 2020 from www.itu.int

ITU-R. (2020e). *Frequently Asked Questions (FAQ) related to Space Plans.* Retrieved September 22, 2020 from www.itu.int

ITU-R. (2003). Resolution 229: Use of the Bands 5 150–5250 MHz, 5250–5350 MHz and 5470–5725 MHz by the mobile service for the implementation of wireless access systems including radio local area networks. In *Provisional Final Acts—World Radiocommunication Conference (WRC-2003).* ITU.

ITU-R. (2015a). *Handbook on national spectrum management.* ITU, Geneva.

ITU-R. (2015b). ITU-R Report SM.2015: Methods for determining national long-term strategies for spectrum utilization. In *ITU-R Reports SM-Seires.* ITU.

ITU-R. (2015c). Resolution 64: Guidelines for the management of unauthorized operation of earth station terminals. In *ITU-R Resolutions.* ITU.

ITU-R. (2015d). *Resolution 958: Urgent studies required in preparation for the 2019 World Radiocommunication Conference* WRC-15 Resolutions. ITU.

ITU-R. (2017). Rules related to satellite systems submitted by an administration acting on behalf of a group of named administrations. In *Rules of procedure approved by the Radio Regulations Board*. ITU.

ITU-R. (2018a). *ITU-R Report SM.2012–6. Economic aspects of spectrum management*. ITU.

ITU-R. (2018b). *Report SM.2093: Guidance on the regulatory framework for national spectrum management*, SM Report Series. ITU.

ITU-R. (2019). *Resolution 22: Measures to limit unauthorized uplink transmissions from earth stations* WRC-19, Sharm El Shiekh.

ITU-R. (2020b). Article 4: Assignment and use of frequencies. In *Radio Regulations*. ITU.

ITU-R. (2020c). Article 11: Notification and recording of frequency assignments. In *Radio Regulations*. ITU, Geneva.

ITU-R. (2020d). Article 18: Licenses. In *Radio Regulations*. ITU.

ITU-R. (2020a). Article 1: Terms and definitions. In *Radio Regulations*. ITU.

Lehr, W. (2004). Dedicated lower-frequency unlicensed spectrum: the economic case for dedicated unlicensed spectrum below 3 GHz. In *New America Foundation, Spectrum Policy Program,Spectrum Series Working Paper 9*.

Lehr, W., & Crowcroft, J. (2005). Managing shared access to a spectrum commons. In *IEEE Symposium on New Frontiers in Dynamic Spectrum Access Networks*, Baltimore.

Lehr, W. (2005). *The role of unlicensed in spectrum reform*. Massachusetts Institute of Technology, USA.

Levin, H. (1988). Emergent markets for orbit spectrum assignments: An idea whose time has come. *Telecommunications Policy, 12*(1), 57–76.

LS telcom AG, Policy Tracker, & Valdani Vicari & Associati. (2017). *Study on spectrum assignment in the European Union*. Retrieved from https://op.europa.eu

Mueller, M. (1982). Property rights in radio communication: The key to the reform of telecommunications regulation. *Cato Policy Analysis* (11).

Noam, E. (1995). Taking the next step beyond spectrum auctions—open spectrum access. *IEEE Communications Magazine, 33*(12), 66–73.

Nozdrin, V. (2008). *Advanced methods of spectrum management for space satellite systems*. Retrieved from www.itu.int

OECD. (2006). *The spectrum dividend: Spectrum management issues*. Retrieved from www.oecd.org

Pavliouk, A. P. (2000). *Incentive radio license fee calculation model*. Retrieved from www.itu.int

Pogorel, G., & Bohlin, E. (2017). *Spectrum 5.0: Improving assignment procedures to meet economic and social policy goals* www.researchgate.net

Roberti, L. (2019). Spectrum policy for satellite and 5G systems: Focus on 28 GHz band. *A KGEC International Journal of Technology, 9*(2), 1–4.

RSPG. (2009). *RSPG report on assignment and pricing methods*. Retrieved from https://rspg-spectrum.eu

Shannon, C. E. (1948). A mathematical theory of communication. *The Bell System Technical Journal, 27*, 379–423.

Smith-NERA. (1996). *Study into the use of spectrum pricing, report prepared by the smith group and NERA for the radio agency*. Retrieved from www.ofcom.org.uk

Vuojala, H., Mustonen, M., Chen, X., Kujanpää, K., Ruuska, P., Höyhtyä, M., Matinmikko-Blue, M., Kalliovaara, J., Talmola, P., & Nyström, A.-G. (2020). Spectrum access options for vertical network service providers in 5G. *Telecommunications Policy, 44*(4).

Weiss, M. B. H., Werbach, K., Sicker, D. C., & Bastidas, C. E. C. (2019). On the application of blockchains to spectrum management. *IEEE Transactions on Cognitive Communications and Networking, 5*(2), 193–205.

Werbach, K. (2004). Supercommons: Toward a unified theory of wireless communication. *Texas Law Review, 82*, 863–973.

Werbach, K. (2010). Castle in the air: A domain name system for spectrum. *Northwestern University Law Review, 104*(2), 613–640.

WIK. (2006). *Towards more flexible spectrum regulation.* Study Commissioned by the German Federal Network Agency. Retrieved 30–5–2019, from http://www.wik.org

Youell, T. (2019). US regulator sells lower 26 GHz band at higher limits. *PolicyTracker*. Retrieved May 30, 2019, from http://www.policytracker.com

Youssef, A., Kalman, E., & Benzoni, L. (1995). Technico-economic methods for radio spectrum assignment. *IEEE Communications Magazine*, *June*, 88–94.

Chapter 8
Developing Countries in the ITU-R

It must be asked whether the increasing emphasis on the needs/requirements of the developing countries might unacceptably distort the functioning of what remains fundamentally an organization dedicated to the technicalities of the international telecommunications.

Francis Lyall (2011)

8.1 Introduction

The presence of developing countries in the ITU-R has evolved over time. Prior to 1950, the ITU was controlled by a small number of European countries, namely France, the UK, Italy and Portugal, that directed the votes of their colonies in Africa, Asia and elsewhere. The number of colonies decreased over time so that by WARC-79, there were no colonies present at the conference. In parallel, the number of developing countries started to increase until they became a majority of Member States. However, the ITU-R RR were perceived by the developing countries at that time as not being relevant to their needs. As a consequence, the developing countries have attempted to change the focus of these regulations away from the technological needs of the industrial world to the demands of least developed countries. They have also demanded preferential treatment for themselves over developed countries.

While the attendance and contributions of developing countries within the ITU-R have not significantly changed over the years, what has changed is the role of leading developing countries and regional organization, and the realization that the major advantage of these countries is in their collective number. They have, therefore, started in recent years to establish regional groups on spectrum management issues in order to have common positions on the different WRCs agenda items and to use their numbers advantageously. Although some have sought to investigate radio spectrum policy in developing countries (El-Moghazi et al., 2008; Wellenius & Neto, 2005, 2007), few have addressed the interaction of a country's policy with the ITU-R radio regulations (El-Moghazi et al., 2015; Jakhu, 2000).

M. A. El-Moghazi and J. Whalley, *The International Radio Regulations*,
https://doi.org/10.1007/978-3-030-88571-7_8

With this is in mind, this chapter's focus is on developing countries and their activities within the ITU-R. It starts with an overview of spectrum management issues in general in developing countries, and then it addresses in more detail the case of the developing countries in the ITU-R with a focus on four aspects. The first two are the resistance and support of the developing countries to new services and technologies, while the third aspect is the importance of the concept of a priori planning to developing countries. The fourth aspect that we shall examine highlights the role of developing countries in WRC-12 with regard to the additional mobile allocation in the 700 MHz band. The chapter then explores ITU-D activities and projects with regard to spectrum management. We also discuss ITU-D Resolution 9 that addresses spectrum management in developing countries, before highlighting those ITU-D spectrum management projects that are in developing countries. A summary is provided in the last section of the chapter.

8.2 Spectrum Management in Developing Countries

It is not a surprise that most of the literature focuses on spectrum management in developed countries. One reason for this is that most of the scholars in the field are from the developed world, while another is the lack of data on spectrum management in developing countries. Thirdly, most of the recent innovations in spectrum management have emerged from countries such as the USA or the UK.

In recent years, however, many developing countries have undertaken steps to reform their ICT policies, including liberalizing the market and privatizing the incumbent operator. Even with the positive changes that these policies have brought about, it is clear that as the telecommunications industry has developed, with new technologies emerging and services launched, a digital divide has emerged between developed and developing countries in terms of access to ICT services. In fact, a closer look at developing countries reveals that most of them depend on wireless technologies due to the lack of alternatives wireline platforms. As a result, the availability of suitable radio spectrum becomes even more critical for these countries. As argued by El-Moghazi et al. (2013, 2015), there should be more focus on what they call the 'spectrum divide' where radio spectrum, which is a critical component for providing wireless services, is typically unavailable or limited in developing countries. The spectrum divide, it is argued, is a direct consequence of the inefficient traditional spectrum management approach ('command and control') that was previously addressed in Chapter Two. Such an approach focuses mostly on technical aspects of spectrum management such as distributing the users in a way that does not lead to harmful interference. However, it ignores other economic, social and functional aspects of spectrum management (Burns, 2002; Wellenius & Neto, 2005). In addition, there is no incentive for current users to return unused spectrum or using new technologies that require smaller spectrum bands (Wellenius & Neto, 2005).

To this end, there are few, yet valuable, explorations in the literature of spectrum management in developing countries. A survey on them shows some deficiencies commonly found in the spectrum policies of developing countries (Cave et al., 2007; El-Moghazi et al., 2008; Wellenius & Neto, 2007). Firstly, spectrum policy is non-transparent and unpredictable which discourages investment, and there is no accurate database of current users or capabilities to monitor and enforce compliance. Secondly, the spectrum license for new services is initially assigned to only one operator and then gradually to other operators which results in artificial scarcity and higher license fees. Spectrum policy is also usually driven largely by political decisions with technology innovation playing a secondary role. For instance, while GSM was introduced in many developed countries in the 1980s, it was postponed in some developing countries until the late 1990s as only then did it gain sufficient political support. Wi-Fi, in contrast, was driven mainly by technology innovation which forced the regulators of developing countries to evacuate the 2.4 GHz spectrum band for the service. Spectrum is considered as a source of revenue by many governments, and the regulatory authority has limited assets in terms of finance, equipment and human resources. Finally, existing users, especially those in the public sector and military, are facing few, if any, incentives to efficiently utilize the spectrum or invest in spectrally efficient technologies as in most cases they are excluded from license fees.

To tackle these deficiencies, a limited literature has emerged. Wellenius and Neto (2005) proposed measures that could be used for enhancing the command and control regime in developing countries such as deploying spectrum refarming, using spectrum sharing, extending the upper end of the spectrum band, introducing auctions and charging government for their spectrum usage. Wellenius and Neto (2007) continued their analysis and designed a framework for reforming the spectrum policy of developing countries that has three main components, namely improving traditional government administration, the establishment of tradable spectrum rights and the development of spectrum commons. They also proposed three steps to establish tradable spectrum rights: defining spectrum rights and obligations, managing interference, and safeguarding fair competition. One of the limitations of the Wellenius & Neto framework, however, is that it only considers two regimes, trading and commons, for the reform process as the solution and it does not address in detail the balance between them.

In practice, Guatemala and El Salvador liberalized their spectrum use in, respectively, 1996 and 1997 via different means, but in both cases it seems that the license holder can deviate from current service allocation as long as it does not cause interference while applying technological neutrality (Hazlett & Muñoz, 2006). A preliminary assessment of these two experiences shows a productive use of the spectrum and the lowering of mobile phone rates, with interference being relatively rare (Ibarguen, 2003) Moreover, it is argued that trading in Guatemala and El Salvador has proven economically efficient in terms of competition and transaction costs. However, these two countries cannot be taken as a successful model for developing countries (Hazlett et al., 2006). Guatemala faces issues such as spectrum hoarding and difficulties retrieving spectrum so that it can be reallocated for license exempt use (Wellenius &

Neto, 2007). El Salvador is also a special case in that it has a small population (of almost seven million) and limited area (20,000 km^2). In addition, Cave et al. (2007) point out that in many developing countries most of the spectrum is not yet assigned or is assigned wastefully to the public sector. Hence, if implementing trading would cause delays in assigning the spectrum, the traditional command and control should be applied instead.

El-Moghazi et al. (2008) state that spectrum trading in developing countries may face many obstacles including government security concerns, weaknesses of the enforcement system and resistance to the concept of spectrum liberalization by stakeholders. They argue that spectrum commons may be considered more appropriate for developing countries as it would reduce scarcity and lower transaction costs and windfall profits. El-Moghazi et al. (2008) also suggest that the ITU activities regarding cognitive radio and software-defined radio would encourage developing countries to adopt the spectrum commons regime.

In general, it is important to address the case of developing countries when it comes to spectrum management reform for a variety of different reasons. Aside from the fact that wireless technologies are quite important for the developing countries due to the absence of suitable alternative wireline solution, one other reason is that these countries play a significant role in international spectrum management due to their active participation in the ITU (Wellenius & Neto, 2005). In addition, Petrazzini (1995) argues that in a relatively closed political system in which the state enjoys considerable independence from civil society and has more executive power, telecommunications policy reform is more successful than in an open political system. Therefore, telecommunications policy reform in developing countries is usually driven by the state without the intervention of civil society or industry.

Rouvinen (2004) argues that the market development of digital technologies has been driven by the needs of developed countries. However, this has recently changed due to two reasons. Firstly, the markets in most developed countries are saturated so manufactures and operators start to focus more on developing countries. Secondly, the expanded user base in developing countries has encouraged the development of local technical and non-technical innovations. Accordingly, developing countries are a suitable laboratory for experimenting new wireless technologies and introducing innovations in national spectrum policy.

8.3 Developing Countries in the ITU-R

The ITU has the greatest developing country involvement in terms of number of countries, dedicated budget to support work related to the developing countries, and dedicated groups with focus on developing issues (MacLean et al., 2002). The work of the ITU with respect to developing countries started in 1952 through the ITU participation in the UN Expanded Programme of Technical Assistance to recruit and send experts to developing countries. This programme subsequently became the United Nations Development Programme 'UNDP'. The second major step was in

1982 when the ITU PP, held in Nairobi, established the Independent Commission for World-Wide Telecommunications Development, which produced the seminal 'The Missing Link' report in 1985. This report highlighted the imbalance in access to telecommunications between developed and developing countries. At the Nairobi PP, several modifications were introduced into ITU formal documents to emphasize the specific needs of developing countries (Lyall, 2011). A third step was taken when the ITU held its first World Telecommunication Development Conference (WTDC) in 1985 in Tanzania. Finally, in 1989, the ITU PP established the Centre for Telecommunication Development which was later incorporated into the current development sector of the ITU (ITU-D) in 1991.

There are also several resolutions passed by ITU PP conferences that support developing countries. For instance, Resolution 170 (Guadalajara, 2010) sought to allow sector members from developing countries to participate in the work of the ITU-R and ITU-T, and to set the level of financial contribution for such participation at one-sixteenth of the level for sector members (ITU, 2010). Another example is Resolution 135, which instructs the Telecommunication Development Bureau (BDT) to continue to provide highly qualified technical experts to developing countries, individually and collectively (ITU, 2018a).

Within the ITU-R, the presence of developing countries has changed over time. Decolonization reduced the number of colonies present in the ITU while simultaneously increasing the number of developing countries. As these countries viewed the ITU-RR as not being relevant to their needs (Rutkowski, 1979), they have sought to change the focus of these regulations away from the technological needs of developed countries towards areas that are more relevant to them. The first critique of the developing countries of the international regime was related to the use of the 'first-come first-served' method of registering frequency assignments (Rutkowski, 1979). The concerns expressed were related to the monopolization of satellite spectrum and orbital resources by developed countries (Sung, 2003). Pressure from developing countries resulted in the adoption of a priori plans for maritime and broadcasting-satellite services (Rutkowski, 1979). A priori planning provides guarantee of access to spectrum, with each country submitting its requirements at a world or regional planning conference (Ryan, 2005).

In addition, developing countries have demanded preferential treatment for themselves over developed countries. Accordingly, the ITU Nairobi Plenipotentiary of 1982 acted to make the provision of technical assistance the primary purpose of the ITU (Codding, 1991). Furthermore, the ITU constitution states explicitly that the usage of the spectrum by ITU Member States should take into account the special needs of developing countries.

Lyall (2011) argues that the insistence of developing countries on their rights may eventually damage the ITU as the increasing emphasis on their requirements could distort the functioning of the organization which focuses on the technical issues of international telecommunications. In other words, there are concerns that the emphasis towards developing countries may contradict with what is supposed to be discussions based on the purely technical aspects of radio spectrum and associated wireless technologies.

Within the RR, there are several parts where developing countries are explicitly mentioned and their needs addressed. Most importantly, the preamble of the RR states that '*in using frequency bands for radio services, Members shall bear in mind that radio frequencies and any associated orbits, including the geostationary-satellite orbit, are limited natural resources and that they must be used rationally, efficiently and economically, in conformity with the provisions of these regulations, so that countries or groups of countries may have equitable access to those orbits and frequencies, taking into account the special needs of the developing countries and the geographical situation of particular countries*'.

Another example of where there is a specific provision for developing countries is Appendix 30, which is related to the planning of broadcasting-satellite service in the 11.7–12.2 GHz (in Region 3), 11.7–12.5 GHz (in Region 1), and 12.2–12.7 GHz (in Region 2) frequency bands (ITU-R, 2020). More specifically, Article 4 states that '*where the proposed assignment involves developing countries, administrations shall seek all practicable solutions conducive to the economic development of the broadcasting-satellite systems of these countries*'.

Other examples are provided in Resolution 5 ('Technical cooperation with the developing countries in the study of propagation in tropical and similar areas') and Resolution 20 ('Technical cooperation with developing countries in the field of aeronautical telecommunications'). Furthermore, Resolution 804, which sets out the principles for establishing agendas for WRCs, notes that there is a need to limit the agenda of conferences, while taking into account of the needs of developing countries in a manner that allows the major issues to be dealt with equitably and efficiently (ITU-R, 2012a).

Regarding IMT, there is a specific directive from the ITU-R Assembly held in 2019 to study the optimal technical and operational characteristics for IMT to meet the needs of developing countries to achieve cost-effective broadband access to global telecommunication networks (ITU-R, 2019). Similarly, ITU-D has previously examined the transition of existing mobile networks to IMT-2000 for developing countries (ITU-D, 2004).

Unfortunately, it has been suggested that most developing countries lack sufficient knowledgeable and experienced staff to participate efficiently in decision-making processes (McCormick, 2007). There are several reasons for this. Firstly, WRCs are conducted for a period of four weeks and issues are discussed in parallel. This requires a huge number of delegates to follow all of the meetings, which is beyond the capabilities of most of developing countries (Jakhu, 2000). Secondly, the scope of the WRCs agenda is usually so extensive that developing countries prefer to focus on just a few issues that are a priority for them (Contant & Warren, 2003). Thirdly, WRCs decisions are largely dependent on the technical studies conducted in the meetings of ITU-R WPs and SGs in the period prior to WRCs. Unlike WRCs, which are conducted just once every four years, WPs and SGs meetings are usually conducted once or twice each year and some of them are held outside Geneva. This makes it difficult for developing countries to attend.

While the attendance and contributions of developing countries have not changed significantly, what has changed is the role of leading developing countries and regional organizations and the realization that the major competitive advantage of the developing countries is their collective numbers (Office of Technology Assessment, 1982). Therefore, developing countries have in recent years started to establish regional groups on spectrum management issue to have common positions on the different WRCs agenda items and to use their voting bloc advantageously. For instance, Arab countries established the Arab Spectrum Management Group (ASMG) in 2001 and African countries formed The African Telecommunication Union (ATU) as part of the African Union (AU) in 2001 (ITU-R, 2010; McCormick, 2005). This has enabled an increasing and different role of developing countries in the ITU-R as will be shown in the following subsections.

8.3.1 Resistance to New Services and Technologies

In this section, we will discuss several cases where developing countries resisted new wireless services and technologies that required unnecessary changes to the RR. One example is the opposition of Arab countries to the European proposal in WRC-03 to develop measures to facilitate the use of common, worldwide frequency bands for implementing terrestrial wireless interactive multimedia applications (TWIM) (CEPT, 2003). The Arab countries argued that no regulatory impediments had been identified, and therefore, there is no need to have such an allocation (ARB, 2003). A related issue to TWIM, which was discussed at WRC-12, was the enhancement of the international spectrum regulatory framework to address the convergence of radiocommunication services (Informal Working Group 3 (IWG-3) of the WRC-07 Advisory Committee (WAC), 2006; ITU-R, 2007b). The ITU-R studies approached the issue from two perspectives. The first focused only on convergence between fixed and mobile services, while the second addressed spectrum allocation issues more generally (ITU-R, 2011). During WRC-12, the Arab countries, as well as other developing countries, called for retaining the current practice with regard to spectrum allocation principles as there was sufficient flexibility within the existing regulatory framework and that the WRC process does not impede the introduction of new technologies (APT, 2011; ASMG, 2011a; Colombia, 2011; Mexico, 2011).

In other cases, developing countries have refused to accept the introduction of particular services or technologies that may contradict the RR. For instance, at WRC-03, the Arab countries opposed the allocation of the 14–14.5 GHz band to aeronautical mobile-satellite service (AMSS). Such an allocation was supported by most of the attendants and promoted by Boeing, which planned to use fixed-satellite services (FSS) satellites to provide high-speed Internet. The Arab countries argued that FSS space stations can only communicate with earth stations that are located at fixed points on the ground according to the ITU-R definition of FSS. Thus, allocating the band to mobile services such as AMSS would result in regulatory inconsistency (Sung, 2003). The discussion of earth stations on board vessels (ESVs) at WRC-03 provides another example of the resistance of developing countries to new services that are not in conformity with the RR. The Arab countries refused to accommodate the use of such service within the FSS, and instead proposed to regulate it as a mobile service as it operated as satellite earth stations on moving ships (Sung, 2003).

One explanation of the resistance of developing countries to new services and technologies is related to the differences between the need of developed countries to introduce new wireless technologies and the concern of developing countries regarding changing their existing equipment (Sung, 1992). In other words, developing countries may object to advanced technically efficient technologies proposed by developed countries if these technologies are not economically efficient to them (Office of Technology Assessment, 1982). Another reason for the confrontation between the two groups of countries is that developing countries consider that the technical studies performed by the ITU-R study groups prior to WRCs are irrelevant to their needs as they were conducted by developed countries (Savage, 1989). This enables Member States to raise objections on the basis of the technical studies being inconclusive.

Developing countries could use ITU regulations to erect barriers to the introduction of new technologies in order to defend their interests and, in cases where there is a threat to their sovereignty, control communications near and over their territories (Sung, 2003). Moreover, developing countries are not facing the same challenges as developed countries, such as the acute scarcity of spectrum and pressure from the manufacturing sector to develop markets for new technologies (Wellenius & Neto, 2005).

One other explanation for the resistance of developing countries to changes to the RR is that these regulations are considered the foundation for their own spectrum policy. Hence, they are concerned that any changes to them may cause difficulties in managing the spectrum nationally. For instance, Columbia called for retaining the existing RR radiocommunication services definition as changes to them, especially the fixed and mobile services, may result in conflicts (Colombia, 2011). In addition, it is argued that in many developing countries a lot of inefficient assignments were made before the telecommunications sector was reformed when there was no independent entity for managing spectrum. It would, therefore, be quite difficult to refarm spectrum in response to the advances in services and technologies (El-Moghazi et al., 2008).

It should be noted that in some cases the objection to issues by developing countries is temporary. More specifically, these countries use their bloc voting capabilities on issues that do not directly affect them so that they can subsequently negotiate on matters that they are concerned about (Office of Technology Assessment, 1982). In other words, the objection to some issues and the attempt to delay the resolution of the issue to the last few days of the conference are related to the resolution of other issues (United States Department of State, 2003). For example, the Arab countries opposed having a primary allocation to the space research service (earth-to-space) in WRC-12. They also opposed the allocation for meteorological aids service in the band below 9 in WRC-12. In both cases, the Arab countries changed their positions (Ofcom, 2012).

8.3.2 Support for New Services and Technologies

While the previous section has shown several examples of resistance to new technologies and services for reasons such as their contradiction with the ITU-R RR, this has not always been the case. For instance, the Arab countries, along with several African countries, supported the European proposal in WRC-03 to have the primary global spectrum allocation for WLAN in the 5 GHz band as a mobile service (ARB, 2003; Kenya, Rwanda, & Tanzania, 2003). This is despite WLAN providing fixed broadband access to the end-user, as it was viewed that having a mobile allocation for such technology may not be in conformity with the RR (Radiocommunication Bureau, 2007).

Moreover, one of the issues that gained support from developing countries is related to the standardization process of cellular mobile standards and IMT process within the ITU—see Chap. 5. In particular, there have been five radio interfaces recommended by the ITU to be the terrestrial component of IMT-2000: CDMA direct spread (WCDMA), CDMA multicarrier (CDMA, 2000), CDMA TDD (TD-SCDMA), TDMA single-carrier (EDGE) and FDMA/TDMA (DECT) (ITU-R, 2009). In 2006, the IEEE proposed a new terrestrial radio interface for inclusion in the IMT-2000 standards, that is, IP-OFDMA.

Several developing countries participated in supporting the inclusion of new technologies into the IMT radio interface. For instance, Egypt, Guatemala, Mexico, Pakistan and Brazil made several contributions to the 2007 ITU-R Radio Assembly (RA-07) supporting the inclusion of WiMax into the radio interfaces approved by the ITU (Brazil, 2007; Egypt, 2007; Guatemala, 2007; Mexico, 2007; Pakistan, 2007). Moreover, Egypt, Columbia, Lebanon, Zimbabwe and the UAE supported the inclusion of the TDD component of IMT-2000 CDMA MC and the FDD component of IMT-2000 OFDM TDD WMAN (WiMax) into the IMT-2000 radio interfaces without it being considered as a new submission (Colombia, 2009; Egypt, 2009; Lebanon, 2009; UAE, 2009; Zimbabwe, 2009).

It can be argued that the support of developing countries for the inclusion of a particular technology into the IMT standards was not support for the technologies. Instead, it should be interpreted as supporting the equal treatment within ITU-R of all technologies as long as it fulfils IMT requirements. Moreover, unlike European countries, which previously mandated particular technologies such as GSM and UMTS in specific technologies, several developing countries supported technology neutrality and, therefore, promoted diversity in technological standards to provide more flexibility to their national operators. In addition, as the private sector companies cannot vote in WRCs and national regulators are the ultimate decision-makers (McCormick, 2007), these companies lobby ITU-R countries to obtain support for their interests when it comes to the drafting of the WRC which they can only attend as observers (Irion, 2009). Therefore, it would appear that developing countries are sometimes used as a proxy by industry to support their particular positions and interests.

In some other cases, developing countries supported the study of new technologies within the ITU-R to obtain some form of international regulations for these technologies under the umbrella of the ITU-R. This would help ensure the protection of existing services from harmful interference. In other words, as it is difficult for developing countries to resist the introduction of new technologies manufactured by industrial countries, they prefer to have ITU-R recommendations for the usage of these technologies instead of developing usage regulations by themselves. For instance, the Arab countries supported the primary global spectrum allocation for WLAN in the 5 GHz band as a mobile service in WRC-03 as it was associated with deploying technical criteria to ensure the protection of fixed and mobile services (ARB, 2003).

Finally, the support to study a particular technology or service could be related to the concerns about the interference caused by technologies imported from developed countries. For instance, the Arab countries proposed WRC-12 Agenda Item 1.22 to examine the effect of emissions from short-range devices on radiocommunication services due to the possibility of equipment to easily move across borders (Ofcom, 2012). The Arab countries also proposed to study regulatory measures and spectrum requirements to enable the introduction of software-defined radio (SDR) and cognitive radio systems (CRS) (ARB, 2007). The Arab countries needed the ITU-R study groups to clarify and regulate for them the operations of these technologies (Billquist, 2008). Another example of the importance of ITU-R regulations on technologies is the survey that was conducted by African countries on the regulation of Wi-Fi. It was found that most African countries refer their national Wi-Fi regulations to ITU-R recommendations even though these recommendations are vague (Neto, 2004). Neto (2004) explains that the African countries want to have a justification for the regulatory choices they have made.

8.3.3 A Priori Planning

Another concept that is relevant to developing countries is the pre-engineering of spectrum, which is often referred to as a priori planning. This was previously discussed in Chapter Four where we demonstrated the difference between a prior planning and first come first served. Here we emphasize the relative importance of the concept for developing countries while focusing on broadcasting and satellite services. In general, a priori planning of frequencies was perceived to be important for developing countries as it achieved equality among different countries. With respect to satellite orbital slots, developing countries managed to obtain several international statements emphasizing the equitable distribution of orbital slots and frequencies for space communications (Zacher, 1996).

The continuous search of developing countries for a priori planning approach was also due to the fear of losing preferred orbital slots and the constraints placed on later comers when it comes to accessing spectrum. On the other hand, developed countries were concerned about the long period of unused orbital slots and barriers for new satellite systems (Levin, 1981). In fact, there were historically several calls for a priori planning the entire spectrum in preference to this occurring incrementally

(Savage, 1989). The idea was dropped due to the fear that developing countries would dominate the decision-making procedures (Zacher, 1996).

First come first served is criticized for not always resulting in the best outcome for the public interest as it may allow commercial satellite operators to occupy a geostationary position that could be better used to provide services for developing countries (Lyall, 2011). Moreover, it is argued that as a priori planning is applied to only a few services and bands, and, therefore, most of the radio spectrum is controlled by those developing countries who registered their frequencies first (Jakhu, 2000). While the first-come first-served concept was criticized by developing countries for keeping most of the radio spectrum controlled by developed countries, countries such as the USA claimed that improvements in technology will permit the expanded use of spectrum which may be restricted by a priori planning (Bortnick, 1981).

The a priori planning concept for orbital satellite slots emerged in the 1970s when developing countries called at the PP of 1973 for a fair distribution of geostationary orbits to restrict the abilities of developed countries to dominate the market (Jakhu, 2000). At that conference, Article 44 was adopted. This called for equitable access to geostationary-satellite orbits and for utilizing them rationally, efficiently and economically. Jakhu (2000) criticized the terms adopted in the article for not being clearly defined.

Another major development with regard to the involvement of developing countries in the satellite industry was in 1976 when eight countries signed the 'Bogotá Declaration' that declared their sovereignty over the geostationary orbit arc above these countries. Eventually, it was not possible to obtain global support for such a declaration (Ryan, 2012). However, a specific phrase was added to the RR pointing to the needs of equatorial countries, namely '*taking into account the special needs of the developing countries and the geographical situation of particular countries*' (Chipman and Jasentuliyana, 1985). At WARC-79 developing countries called for a recognition of their special needs and for an end to such an approach as it puts an inequitable burden on latecomers (Rutkowski, 1985). Following this, there have been several a priori plans (Wang, 2018). For instance, WARC Orbit Conference (ORB-88) planned to provide equitable and guaranteed access by all countries to the geostationary-satellite orbit (GSO) (ITU, 1988).

With respect to broadcasting services, there have been several Regional Radiocommunication Conferences in the past providing developing countries with the opportunity to organize the assignment of such services. For instance, the ITU held the African YHF/UHF Broadcasting Conference in 1963 to plan for the assignment of frequencies for national broadcasting and television stations in the VHF and UHF bands across Africa (ITU, 1963). These Regional Radiocommunication Conferences are regulated by Article 9 of the ITU Convention, which states that '*[t]he agenda of a Regional Radiocommunication Conference may provide only for specific radiocommunication questions of a regional nature, including instructions to the Radio Regulations Board and the Radiocommunication Bureau regarding their activities in respect of the region concerned, provided such instructions do not conflict with the interests of other regions. Only items included in its agenda may be discussed by such a conference*'.

With regard to analogue terrestrial broadcasting in Region 1, the service was originally covered in Europe by what is called the Stockholm Plan (ST61), which was developed in 1961, and in African and Arab countries by the Geneva Plan (GE89) that was established in 1989 (Beutler, 2008). CEPT countries requested that the ITU conduct a Regional Radiocommunication Conference (RRC) for the revision of the ST61 plan to introduce digital broadcasting across the European Broadcasting Area (EBA). Due to the interest of African countries, as well as their neighbouring countries, it was decided to convene a RRC for the planning of digital terrestrial broadcasting services in Region 1 and in the Islamic Republic of Iran, in the 174–230 MHz and 470–862 MHz bands (Beutler, 2008). That plan superseded the ST61 in Europe and GE89 in the African Broadcasting Area (ABA).

Ultimately, the ITU Regional Radio Conference (RRC-06), which was held in two sessions in 2004 and 2006, planned the process of the digital switchover of terrestrial broadcasting services in Europe, Africa and the Middle East (Irion, 2009). RRC-06 resulted in the Geneve-2006 (GE-06) agreement that planned the analogue and digital broadcasting services in the 174–230 MHz and 470–862 MHz bands. The GE-06 plan determined the end of the transition period for analogue broadcasting services in the UHF band to be 17th June 2015 (GSMA, 2012).

The discussion during RRC-06 regarding the transition period, after which analogue television stations would not be protected, revealed the difference in requirements between CEPT countries on the one hand and developing countries in Africa and the Arab world on the other. More specifically, European countries wanted to switch to digital TV as soon as possible while there was a view among African and Arab countries of not switching off analogue stations before 2028 or after 2038. Eventually, it was agreed to end the transition period by 2015 except for the VHF band where some countries received an extension until 2020 (Beutler, 2009). It is argued that in response to the disagreement over the transition period, the envelope concept was introduced into the plan where an analogue station can operate under the envelope of a digital plan entry (Beutler, 2009).

8.3.4 *The 700 MHz Mobile Allocation in WRC-12*

While the previous three subsections have recounted examples of support for, or resistance to, new technologies, services or the preplanning of spectrum, none of them were solely initiated by developing countries. Instead, these positions were in response to proposals made by developed countries. However, the discussion regarding the additional allocation of mobile services in the 700 MHz band at the WRC-12 witnessed a significant change in the stance of developing countries. Quite simply, they become more proactive than had previously been the case.

Historically, most of the UHF band (470–862 MHz) had been allocated to analogue terrestrial broadcasting services in Region 1 for many decades. The first change to such a plan was at the 2006 Regional Radiocommunication Conference (RRC-06), which planned the digital terrestrial broadcasting service in Region 1 and the Islamic

Republic of Iran in the 174–230 MHz and 470–862 MHz frequency bands (ITU, 2006). Shortly afterwards, pressure from the mobile telecommunications industry motivated the WRC-07 to make an additional allocation in the 790–862 MHz frequency band to mobile services effective from June 2015 (ITU-R, 2007a). The conference also resolved to invite the ITU-R to conduct sharing studies for Regions 1 and 3 in the 790–862 MHz band between mobile and other services in order to protect the services such as broadcasting to which the frequency band had already allocated (ITU-R, 2007c). Subsequent to these sharing studies, it was agreed at WRC-12 that no new mandatory regulatory measures were needed to enable sharing between mobile and broadcasting services in neighbouring countries in the 790–862 MHz band (Ofcom, 2012).

The Arab and African countries called during WRC-12 for the 694–790 MHz band to be immediately allocated to mobile services in Region 1, as it was in ITU Regions 2 and 3, to meet the growing demand for broadband (Standeford, 2012a, 2012b). European countries opposed such a proposal because the band was mainly allocated for broadcasting service in their territories and a large investment had already been made to fund the transition to digital television (Sims, 2012). WRC-12 eventually decided to allocate the 694–790 MHz band in Region 1 to mobile service on a primary basis in addition to the existing primary broadcasting service. The allocation was effective after WRC-15 dependent on the refinement of the lower edge of the allocation (ITU-R, 2012b). It is worth mentioning that several European countries recorded their reservations and stated that they agreed to reach a compromise with great reluctance, on an exceptional basis in the spirit of international cooperation and to satisfy the urgent demands of African and Arab countries. The European countries also stated that WRC-12 neither discussed nor clarified whether the proposal of the 700 MHz belonged to one of the agenda items of WRC-12 (Oberst, 2012).

Not only did WRC-12 witness proactivity from developing countries, but also changes in their positions from previous WRCs. Firstly, the Arab and African countries were united against introducing non-broadcasting services in the spectrum planned for digital TV during the RRC-06 conference (O'Leary et al., 2006). During RRC-06, most Arab countries were not considering allocating mobile services to broadcasting spectrum in the UHF band (Sims, 2007). Following RRC-06, WRC-07 discussed having an additional allocation in the 790–862 MHz band for mobile service in addition to broadcasting services (Beutler, 2012). The Arab countries were divided at WRC-07, with some like Egypt strongly supporting the issue while others, like the UAE, were not in favour of such an allocation (Bateson, 2009). The African countries supported having the 806–862 MHz band allocated to mobile service at WRC-07 (ATU, 2007). However, they also opposed the allocation of the 470–806 MHz band to mobile services arguing that the band is currently used for analogue television terrestrial broadcasting services and, due to RRC-06, for digital terrestrial television (ATU, 2007). Surprisingly, the positions of the Arab and African countries changed in WRC-12 to support having an additional mobile allocation in the 694–790 MHz band.

Before explaining the proactivity of developing countries in WRC-12 with respect to the 700 MHz issue, we first need to outline in more detail the origins of the issue.

The 700 MHz proposal was discussed and initiated very shortly before WRC-12 commenced in January 2012. It was a proposal from the UAE, made at the end of 2011, which was supported by two Gulf countries, namely Qatar and Kuwait (UAE, Qatar, & Kuwait, 2011). The proposal was promoted by several industry-oriented organizations at the first African Telecommunications Union summit on the digital dividend that was held in November 2011 (Kirkaldy, 2011; Lyons, 2011). The summit recommended to pursue the allocation of the 694–790 MHz band to mobile services on an equal primary basis with broadcasting for African countries during WRC-12 (ATU, 2011b). In addition, the mobile industry lobbied the regional organizations of African and Arab countries prior to WRC-12 to support such a proposal (Billquist, 2010a, 2010b). The Arab countries submitted contributions at WRC-12 supporting the issue and called for the harmonization of allocations in the 698–790 MHz band range across the ITU's three regions (ASMG, 2011b). The African countries also promoted the allocation, explaining that the 790–862 MHz band is partially allocated to other services in many African countries which increases the importance of the 694–790 MHz band (ATU, 2011a).

As shown by the above, unlike the normal procedures of discussing only items which are on the agenda of the conference and that have been examined by one of the ITU-R study groups (ITU-R, 2012a), the 700 MHz issue was initiated by a handful of countries shortly before WRC-12 and then encouraged by Arab and African countries during the conference even though it was not one of the conference agenda items. It is worth mentioning that over the course of recent WRCs, there is only one known example of discussing and approving an issue that is not on the agenda of the conference. This occurred in WRC-95 when the conference allocated 400 MHz in the 19 and 29 GHz spectrum bands to non-geostationary fixed-satellite service networks subject to a large number of reservations (ITU, 1996; ITU-R, 1995). Although the issue was not on the agenda (ITU Council, 1994), the conference decided to discuss the issue under pressure from Teledesic, a US-based operator of low-orbit satellites (Radiocommunications Agency, 1997). What the 700 MHz issue clearly shows is that the developing countries were able to be proactive and overcome some deficiencies of the RR such as the limitation of the discussion to items that are on the agenda of WRCs.

8.4 Spectrum Management in the ITU-D

The activities within the ITU with respect to spectrum management are not limited to the ITU-R. There are several activities within the ITU-D that aim to provide assistance to developing countries with regard to specific spectrum management areas. These areas include frequency planning and assignment, management and monitoring, setting fees for spectrum utilization (ITU-D, 2020e). There are also several projects which the ITU-D has led, such as providing a spectrum management system for Colombia, border coordination in Central African countries, assistance on spectrum pricing for Kenya, and satellite communications assistance to Sri Lanka,

Bangladesh and Mongolia (ITU-D, 2020a). In addition, the ITU-D has developed an incentive radio license fees calculation model (Pavliouk, 2000).

An important contribution to spectrum management in developing countries is the automated technical and administrative tool for spectrum management in developing countries which is called SMS4DC (Spectrum Management System for Developing Countries). SMS4DC was developed by the ITU-D as a low-cost tool for managing land-mobile, fixed and broadcasting services (ITU-D, 2020d). SMS4DC was developed in cooperation with ITU-R BR based on the technical specifications developed by the ITU-R and ITU-D group of experts. Importantly ITU-R Resolution 11 states that the ITU-R Study Group 1 and BR experts should continue to assist in the further development of the SMS4DC in accordance with WRC decisions and relevant ITU-R Recommendations, Handbooks and Reports (ITU-R, 2015). In other words, this work will continue in the future.

Another important contribution of the ITU-D regarding spectrum management is the ITU-D 'ICT Eye' portal. This portal provides updated data on topics such as assigned spectrum for 3G (IMT), assigned spectrum for WiMAX services, assigned spectrum for LTE services, and secondary trading (ITU-D, 2020b). Moreover, the 'Global Symposium for Regulators' (GSR), which is organized by ITU-D, provides updated research on contemporary spectrum issues such as its economic valuation and broadband satellite regulations (Alden, 2011; Horton, 2012).

The areas of focus for the ITU-D are usually selected by World Telecommunication Development Conferences (WTDC). The 2017 WTDC, which was held in Bueno Aries (Argentina), adopted several regional initiatives with a focus on spectrum management. In Africa, there is an initiative that focused on the management and monitoring of radio-frequency spectrum and the transition to digital broadcasting, while in the Americas there is an initiative that addressed spectrum management and the transition to digital broadcasting (ITU-D, 2017a).

One of the areas that have attracted special attention from the ITU-D is the transition from analogue to digital broadcasting, where the Telecommunication Development Bureau (BDT) provides assistance regarding new broadcasting services and allocation of the digital dividend. The ITU-D published in 2010 guidelines for the transition from analogue to digital broadcasting, focusing on Africa and the GE-06 agreement (ITU-D, 2010a). In addition, ITU-D has established a project in Africa to collect relevant information on the current status of television broadcasting as well as the programmes undertaken by African broadcasters to shift Analogue to digital terrestrial TV (DTTB) and mobile TV (MTV), and customizing guidelines for selected African countries (ITU-D, 2020c).

8.4.1 World Telecommunication Development Conference Resolution 9

One of the main activities within the ITU-D with respect to spectrum management is related to Resolution 9 ('Participation of countries, particularly Developing Countries, in frequency spectrum management') of the WTDC. The origin of this resolution can be traced back to the WTDC held in 1998 that focused at that time on reviewing national spectrum management and the use of the spectrum in the 29.7–960 MHz frequency band. Accordingly, ITU-R Study Group 1 and ITU-D established a joint ITU-R/ITU-D group to prepare a report. Following this, the 2002 WTDC focused on the 960–3000 MHz band, while WTDC 2006 addressed the 2900 MHz–30 GHz band (ITU-D, 2006, 2010b).

The fourth report in response to Resolution 9, which was revised at WTDC 2010 and then presented at WTDC-14, focused on issues such as the market mechanisms used for frequency assignment, spectrum refarming and spectrum fees (ITU-D, 2014). The fifth and latest report was presented at WTDC-17. It addressed emerging spectrum management approaches with a focus on spectrum sharing, the economic benefits of using licensed and license exempt spectrum in addition to spectrum monitoring (ITU-D, 2017b).

It is worth highlighting that the approved fifth report was not the same one approved by the ITU-D. The last ITU-D SG1 before WTDC-17 in March 2017 approved the report as prepared by the ITU-D/ITU-R Joint Group. However, the Radiocommunication Advisory Group (RAG) expressed its concern that not all of the modifications requested by ITU-R had been included in the final report. Telecommunication Development Advisory Group (TDAG) agreed that ITU-R SG1 should be given the opportunity to consider the report at its June 2017 meeting in order to reach a consensus before WTDC-17. This occurred, enabling the ITU-D SG1 management team to agree to submit the report, albeit with some amendments, to WTDC-17 as the report for Resolution 9 (Chairman ITU-D Study Group1, 2017).

During the discussions within the ITU-R on the revision of the ITU-D report, several countries expressed their concerns about the report's stance towards new technologies which are unproven. More specifically, reservations were expressed that the report placed considerable emphasis on two particular technologies, namely TVWS and DSA. This resulted in the report being unbalanced and resembling a commercial and marketing document rather than an ITU document (Germany (Federal Republic of) et al., 2017). The ITU-R group responsible of revising the report for Resolution 9 changed its name from 'Emerging spectrum management approaches' to 'Spectrum management approaches', to include several systems and approaches instead of focusing primarily on one approach (DSA) and one technology (TVWS) (Chairman Working Party 1B, 2017).

During WTDC-17, the CEPT countries required some changes to Resolution 9 with the main requirement being to discontinue the production of a specific report on the resolution. This was usually prepared by a handful of people, which compromised the scope, neutrality and consistency of the report (CEPT, 2017). Eventually,

Resolution 9 was amended to reflect a form of cooperation between ITU-D and ITU-R in several areas not related to specific technologies or systems (e.g. TVWS) or that required any change in the RR (ITU-D, 2017c).

The most significant change in these examples is with respect to 'New spectrum-access approaches', which focused in the previous version on the resolution of DSA approaches. In the new version, the section was renamed as 'Emerging technologies and approaches in using spectrum' and focused on a number of general systems and technologies: spectrum sharing (DSS), use of satellite and high-altitude platform systems for the provision of services in remote and inaccessible areas, Internet of things (IoT), IMT-2020 and short-range devices. New areas of spectrum management were also added to the resolution such as innovative ways to license spectrum (e.g. LSA, ASA), as well as assistance with dealing with seasonal interference.

Interestingly, one of the major changes to Resolution 9 was that in the versions approved by previous conferences (WTDC-06, WTDC-10, WTDC-14), the resolution resolved to prepare a report within the next study period on specific topics of interest to developing countries regarding spectrum management. However, the version of Resolution 9 approved by WTDC-17 instructed the Director of the Telecommunication Development Bureau, in consultation with the Director of the Radiocommunication Bureau, to prepare reports on spectrum management and monitoring (ITU-D, 2017c). These reports were to look at the issue from multiple perspectives (i.e. technical, economics, regulatory and financial). In other words, it does not seem that there would be a joint group between the ITU-R and the ITU-D to work on a report in response to Resolution 9 as was the practice in the past.

8.4.2 *ITU Spectrum Management Projects in Developing Countries*

The ITU has several activities with a focus on spectrum management in developing countries, with ITU regional offices typically playing a key role in their development and delivery (ITU, 2020c). It was at the ITU PP of 1982 where the idea of a regional presence of the ITU, to strengthen the technical assistance provided by the ITU to developing countries, first emerged. The ITU PP of 1989 argued that a stronger presence of the ITU in the developing countries was required and asked the SG to carry out studies to strengthen such a presence (Posta & Terzi, 2009). The ITU PP of 1994, which was held in Kyoto (Japan) explicitly adopted Resolution 25, which considered the need for an enhanced regional presence and resolved that the principal aim of the regional presence was to enable the ITU to be as close as possible to its members, especially those that are developing countries (ITU, 1994). It also instructed the Council of the ITU to establish a group to evaluate the regional presence of the ITU. It is worth remembering that while the regional offices are part of BDT, they represent the ITU as a whole (Posta & Terzi, 2009).

Resolution 25 was revised at the PP of 2018, held in Dubai, to cover the concerns of some countries regarding the efficiency of the ITU regional presence. For instance, at the conference, the Arab countries proposed to amend Resolution 25 to direct the Deputy Secretary-General to implement measures to further strengthen the regional presence. These included expanding and strengthening the regional and area offices by identifying functions which could be decentralized and implementing them as soon as possible. They also proposed that the regional offices should report annually to the Member States of the relevant region on their activities (Arab States, 2018).

Eventually, the resolution was significantly amended, not least to ensure that the regional offices acted as an extension of the ITU more generally. More specifically, the revised resolution sought to strengthen the functions of the regional offices so that they can play a part in implementing the ITU strategic plan, programmes, etc., set out by WTDC 2017. It also instructed the Director of the ITU-D BR to effectively incorporate into its regional and area offices the activities of BR and Telecommunication Standardization Bureau (TSB) (ITU, 2018b).

With this in mind, it is worth briefly mentioning three large-scale projects to illustrate the nature and focus of the initiatives that have been undertaken. The first one was established by the ITU in West African countries in 2004, in cooperation with the European Union, to support the establishment of an integrated ICT market across the region. One part of the project was the harmonization of the development of the telecommunications sector in West Africa with a focus on different aspects including the management of scarce resources like spectrum (ITU-D, 2005). The project's final guidelines recommended that Economic Community of West African States (ECOWAS) / West African Economic and Monetary Union (UEMOA) countries should flexibly manage spectrum while respecting the ITU international allocations. It also recommended that the countries should establish a common framework for developing a public register (i.e. a database) of technical and locational information about radio systems (ITU, 2005).

A second illustrative ITU-R project was established in Africa following the allocation of the 700 MHz band to mobile services in Region 1. The project was supported by the European Union, co-operating with the ITU and sought to clear broadcasting services from the 700 MHz band so that it could be utilized for 4G (Standeford, 2018). This was in response to Resolution 750 of WRC-15 that sought to bring assistance to those developing countries wishing to implement the new mobile allocation. A more recent project, 'Policy and Regulation Initiative for Digital Africa' (PRIDA), is supported by the African Union (AU), European Union and ITU to create a harmonized set of rules and regulations for the use of ICT (ITU, 2020a). One of the pillars of PRIDA is the efficient and harmonized spectrum utilization with a focus on issues such as spectrum licensing, spectrum pricing and treatment of harmful interference (ITU, 2020b).

One may wonder why most of the ITU projects in Africa are supported by European countries. This may be related to the nature of the radio spectrum and the fact that both Europe and Africa are in ITU Region 1 and are close geographically. This could explain the desire to have a harmonized spectrum policy in Africa that is aligned with Europe. In particular, when Europe clears broadcasting services from

700 MHz so that it can be utilized for mobile, it is in their interest to have the same band harmonized in Africa to achieve economy of scales and to operate safely without interference. This is also the reason why the RRC-06 plan was developed for European and African countries at the same time.

It has been suggested that the focus of the ITU on developing countries is, in fact, related to the needs of developed countries on the one hand and the nature of the telecommunication industry and radio spectrum on the other. More specifically, the telecommunication industry is based on economies of scales and standardization. Therefore, it is important to align developing with developed countries in terms of the adoption and utilization of technologies. Furthermore, wireless communications require harmonization and protection against interference. It is, therefore, in the interest of developed countries to assist developing countries to ensure they are not utilizing inferior technologies, proprietary standards or frequency arrangement plans that contradict with their objectives.

8.5 Summary

Spectrum management in developing countries is largely based on the ‘command and control’ approach and accommodates several deficiencies including being non-transparent and unpredictable, driven by political decisions and perceiving spectrum as source of revenue. Within the ITU-R, the presence of developing countries has evolved over time from being controlled by a small number of European countries to becoming a majority. The developing countries have also started in recent years to establish regional groups on spectrum management issue to have common positions on the different WRCs agenda items and to use their voting bloc advantageously.

There have been situations where there is resistance to new technologies or services that are inconsistent with the RR or that cause unnecessary changes to the well-established international spectrum management regime. One explanation of such resistance is related to the differences between the needs of the developed countries to introduce wireless technologies and the concerns of the developing countries regarding changing their existing equipment. Regarding the support of particular technologies, this chapter has argued that it was not support to the technologies per se but, instead, it was a support to the equal treatment within ITU-R to all technologies that have motivated developing countries.

Another concept that is relevant to developing countries is the a priori planning which is perceived to be important for these countries as it achieved equality among different countries, especially for satellite orbital slots. With respect to the broadcasting service, there have been several Regional Radiocommunication Conferences in the past providing developing countries with the opportunity to organize the assignment of such services with the RRC-06 being the last one. The discussion regarding the additional allocation of mobile services in the 700 MHz at the WRC-12 witnessed a significant change in the stance of developing countries to be more proactive and to even lead the European countries within Region 1.

Finally, we have shown how spectrum management is not limited to ITU-R as several initiatives are located within ITU-D. One of the main activities within ITU-D is related to Resolution 9 of the WTDC, which relates to the participation of countries, especially developing countries, in spectrum frequency management. ITU regional offices play a key role in the development and delivery of these initiatives.

References

Alden, J. (2011). *Exploring the value and economic valuation of spectrum*. Paper presented at the global symposium for regulators. Armenia. Retrieved from www.itu.int

APT. (2011). *Common proposals for the work of the conference. Agenda item 1.2*. Paper presented at the World Radiocommunication Conference (WRC-12), Geneva.

Arab States. (2018). *Arab States common proposal to the conference*. Paper presented at the PP-2018, Dubai. www.itu.int

ARB. (2003). *Arab states common proposal*. Paper presented at the World Radiocommunications Conference 2003, Geneva. Retrieved from www.itu.int

ARB. (2007). *Arab States common proposals*. Paper presented at the World Radiocommunications Conference 2007, Geneva. Retrieved from www.itu.int

ASMG. (2011a). *Arab states common proposals. Common proposals for the work of the conference. Agenda item 1.2*. Paper presented at the World Radiocommunication Conference (WRC-12), Geneva.

ASMG. (2011b). *Arab states common proposals. Common proposals for the work of the conference. Agenda item 1.17*. Paper presented at the World Radiocommunication Conference (WRC-12), Geneva.

ATU. (2007). *African common proposals for the work of the conference. Agenda item 1.4*. Paper presented at the World Radiocommunication Conference (WRC-07), Geneva.

ATU. (2011a). *African common proposals for the work of the conference. Agenda item 1.17*. Paper presented at the World Radiocommunication Conference (WRC-12), Geneva.

ATU. (2011b). *Recommendations*. Paper presented at the ATU Digital Migration and Spectrum Policy Summit, Nairobi.

Bateson, R. (2009). Opinion—globalisation starts to unpick traditional regional alliances. *Policy-Tracker*. Retrieved from www.policytracker.com

Beutler, R. (2008). The regional radio communication conference RRC-06 and the GE06 agreement. In *Digital terrestrial broadcasting networks*. Springer.

Beutler, R. (2012). *Results of the WRC-12 from a European broadcaster's perspective*. Paper presented at the LS Summit 2012. Retrieved from http://www.lstelcom.com

Beutler, R. (2009). *Digital terrestrial broadcasting networks*. Springer.

Billquist, S. (2008). ITU considers how to stimulate global development of cognitive radio. *PolicyTracker*. Retrieved from www.policytracker.com

Billquist, S. (2010a). European pressure grows for 2016 global IMT spectrum action. *PolicyTracker*. Retrieved from www.policytracker.com

Billquist, S. (2010b). Mobile Industry to Press Spectrum Needs in 2016 World Conference Preparations. *PolicyTracker*. Retrieved from www.policytracker.com

Bortnick, J. (1981). International information flow: The developing world perspective. *Cornell International Law Journal, 14*(2), 333–353.

Brazil. (2007). *View on extension of the recommendation ITU-R M.1457*. Paper presented at the ITU WP 8F Meeting. Retrieved from www.itu.int

Burns, J. (2002). *Measuring spectrum efficiency—the art of spectrum utilization metrics*. Paper presented at the IEE Conference on Getting the Most Out of Spectrum, London, UK.

Cave, M., Doyle, C., & Webb, W. (2007). *Essentials of modern spectrum management. The Cambridge wireless essentials series* (pp. xii, 265 p.).
CEPT. (2003). *WRC-03: European common proposals for the work of the conference*. Geneva.
CEPT. (2017). *Proposal for the work of the conference*. Paper presented at the WTDC-17, Buenos Aires.
Chairman ITU-D Study Group 1. (2017). *Report of ITU-D SG1 to WTDC, world telecommunication development conference 2017 (WTDC-17)*. Paper presented at the WTDC-17, Buenos Aires. Retrieved from www.itu.int
Chairman Working Party 1B. (2017). *Report on the third meeting of WP 1B*. Paper presented at the WP 1B 3rd Meeting, Geneva.
Chipman, R., & Jasentuliyana, N. (1985). Developing countries, the GEO and the WARC-ORB 85 conference. *Space Policy, 1*(3), 244–249.
Codding, G. A. (1991). Evolution of the ITU. *Telecommunications Policy, 15*(4), 271–285.
Colombia. (2009). *View on updated of the recommendation ITU-R M.1457–9*. Paper presented at the ITU WP 5D Meeting. Geneva.
Colombia. (2011). *Proposal for the work of the conference. Agenda item 1.2*. Paper presented at the World Radiocommunication Conference (WRC-12), Geneva.
Contant, C. M., & Warren, J. (2003). The world radiocommunication conferences process: Help or hindrance to new satellite development? *Acta Astronautica, 53*(4).
Egypt. (2007). *View on extension of the recommendation ITU-R M.1457*. Paper presented at the ITU WP 8F Meeting. Retrieved from www.itu.int
Egypt. (2009). *View on revision of the recommendation ITU-R M.1457–8*. Paper presented at the ITU WP 5D Fifth Meeting. Retrieved from www.itu.int
El-Moghazi, M., Digham, F., & Azzouz, E. (2008). *Radio spectrum policy reform in developing countries*. Paper presented at the IEEE Symposium on New Frontiers in Dynamic Spectrum Access Networks, Chicago.
El-Moghazi, M., Whalley, J., & Irvine, J. (2013). *International spectrum management regime: a case of regulatory lock-in for the developing countries?* Paper presented at the CPR Asia/CPR Africa, Mysore.
El-Moghazi, M., Whalley, J., & Irvine, J. (2015). *The 700 MHz mobile allocation in Africa: Observations from the battlefield*. Paper presented at the CPR Asia/CPR Africa, Taiwan.
Germany (Federal Republic of), Saudi Arabia (Kingdom of), Armenia (Republic of), China (People's Republic of), Djibouti (Republic of), Egypt (Arab Republic of), ... of), Z. R. (2017). *Report on WTDC-14 resolution 9 (REV. Dubai, 2014)*. Paper presented at the WTDC-17, Buenos Aires.
GSMA. (2012). *Geneva 06: Regional radio conference*. Retrieved from http://www.gsma.com
Guatemala. (2007). *Views related with proposal in Doc 8f/1065*. Paper presented at the ITU WP 8F Meeting. Retrieved from www.itu.int
Hazlett, T. W., Ibarguen, G., & Leighton, W. A. (2006). *Property rights to radio spectrum in Guatemala and El Salvador: An experiment in liberalization*. Paper presented at the Telecommunications Policy Research Conference, Washington D.C.
Hazlett, T. W., & Muñoz, R. E. (2006). Spectrum allocation in Latin America: An economic analysis. *George Mason Law & Economics Research Paper, 6*(44), 261–278.
Horton, B. (2012). *Spectrum policy in a hyperconnected digital mobile world*. Paper presented at the Global Symposium for Regulators Colombo. Retrieved from www.itu.int
Ibarguen, G. (2003). Liberating the radio spectrum in Guatemala. *Telecommunications Policy, 27*(7), 543–554.
Informal Working Group 3 (IWG-3) of the WRC-07 Advisory Committee (WAC). (2006). *IWG-3 comments on the RCS proposal on agenda item 4 (including Recommendation 722 (WRC-03))*.
Irion, K. (2009). Separated together: The international telecommunications union and civil society. *International Journal of Communications Law and Policy, 13*, 95–113.
ITU. (1963). African broadcasting conference. *Telecommunication Journal, 30*(6), 158. www.itu.int

ITU. (1988). ORB-88 adopts plan and regulatory provisions for geostationary satellites. *Telecommunication Journal, 55*(12), 790–794. www.itu.int

ITU. (1994). *Resolution 25: ITU PP regional presence.* Paper presented at the PP-1994, Kyoto. www.itu.int

ITU Council. (1994). *Agenda for the world radiocommunication conference (WRC-95).* Retrieved from www.itu.int

ITU. (1996). WRC-95: A new approach. *ITU News,* (1). Retrieved from www.itu.int

ITU. (2005). *Harmonization of policies governing the ICT market in the UEMOA-ECOWAS space.* Retrieved from www.itu.int

ITU. (2006). Article 1: Definitions. In *Final acts of the regional radiocommunication conference for planning of the digital terrestrial broadcasting service in parts of regions 1 and 3, in the frequency bands 174–230 MHz and 470–862 MHz (RRC-06).* ITU.

ITU. (2010). *Resolution 170: Admission of sector members from developing countries to participate in the work of the ITU radiocommunication sector and the ITU telecommunication standardization sector.* Paper presented at the PP-10, ITU. Retrieved from www.itu.int

ITU. (2011a). *Convention of the international telecommunication union.* ITU.

ITU. (2011b). *Constitution of the international telecommunication union.* ITU.

ITU. (2018a). *ITU's role in the development of telecommunications/ICTs, in providing technical assistance and advice to developing countries 1 and in implementing relevant national, regional and interregional projects.* Paper presented at the PP-19, Dubai, ITU

ITU. (2018b). *Strengthening the regional presence.* Paper presented at the PP-2018. ITU, Dubai. Retrieved from www.itu.int

ITU. (2020b). *PRIDA: Spectrum harmonization.* Retrieved from https://oneprida.africa

ITU. (2020a). *Policy and regulation initiative for digital Africa (PRIDA).* Retrieved from www.itu.int

ITU. (2020c). *Regional presence.* Retrieved from www.itu.int

ITU-D. (2004). *Mid-term guidelines (MTG) on the smooth transition of existing mobile networks to IMT-2000 for developing countries.* Retrieved from www.itu.int

ITU-D. (2005). *West African common market, harmonization of policies governing the ICT market in the UEMOA-ECOWAS space, project: Radio Spectrum Management.* Retrieved from www.itu.int

ITU-D. (2006). Resolution 9: Participation of Countries, Particularly Developing Countries, in Spectrum Management. In *The World Telecommunication Development Conference (Doha, 2007) Final Act.* ITU.

ITU-D. (2010a). *Guidelines for the transition from analogue to digital broadcasting.* ITU. Retrieved from www.itu.int

ITU-D. (2010b). *Report on resolution 9 (Rev. Doha, 2006): Participation of countries, particularly developing countries in spectrum management.* ITU. Retrieved from www.itu.int

ITU-D. (2014). *Report on resolution 9: Participation of countries, particularly developing countries, in spectrum management.* ITU. Retrieved from www.itu.int

ITU-D. (2017a). *A snapshot—report of the world telecommunication development conference 2017 (WTDC-17)).* ITU. Retrieved from www.itu.int

ITU-D. (2017b). *Resolution 9 Participation of countries, particularly developing countries, in spectrum management: Evolving spectrum management tools to support development needs.* ITU. Retrieved from www.itu.int

ITU-D. (2017c). Resolution 9 (Rev. Buenos Aires, 2017): Participation of countries, particularly developing countries, in spectrum management. In *WDTC-2017 Final Acts.* ITU.

ITU-D. (2020a). *Assistance on spectrum management.* ITU. Retrieved from www.itu.int

ITU-D. (2020b). *ICT eye.* ITU. Retrieved from www.itu.int

ITU-D. (2020c). *Project on the digital broadcasting transition roadmap in Africa.* ITU. Retrieved from www.itu.int

ITU-D. (2020d). *SMS4DC version 4.0.* ITU. Retrieved from www.itu.int

ITU-D. (2020e). *Spectrum management.* ITU. Retrieved from www.itu.int

ITU-R. (1995). WRC-95 Resolution 121. Development of interference criteria and methodologies for coordination between feeder links of non geostationary satellite networks in the mobile satellite service and geostationary satellite networks in the fixed satellite service in the bands 19.3–19.6 GHz and 29.1–29.4 GHz. In *Provisional Final Acts—World Radiocommunication Conference (WRC-95).* ITU. Retrieved from www.itu.int

ITU-R. (2007a). *Article 5: Frequency allocations.* In Radio Regulations. ITU. Retrieved from www.itu.int

ITU-R. (2007b). Resolution 805. Agenda for the 2011 world radiocommunication conference. In *Provisional Final Acts—World Radiocommunication Conference (WRC-07).* ITU.

ITU-R. (2007c). WRC-07 Resolution 749: Studies on the use of the band 790–862 MHz by mobile applications and by other services. In *Provisional Final Acts—World Radiocommunication Conference (WRC-07).* ITU. Retrieved from www.itu.int

ITU-R. (2009). ITU-R Recommendation M.1457–8: Detailed specifications of the radio interfaces of international mobile telecommunications-2000 (IMT-2000). In *M Series. Mobile, Radiodetermination Amateur and Related Satellite Services.* ITU

ITU-R. (2010). *ITU-R report SM.2093–1: Guidance on the regulatory framework for national spectrum management.* ITU. Retrieved from www.itu.int

ITU-R. (2011). *CPM report on technical, operational and regulatory/procedural matters to be considered by the 2012 world radiocommunication conference.* Geneva.

ITU-R. (2012a). WRC-12 Resolution 804: Principles for establishing agendas for world radiocommunication conferences. In *Provisional Final Acts—World Radiocommunication Conference (WRC-12).* ITU.

ITU-R. (2012b). WRC-12 Resolution 232. Use of the frequency 694–790 MHz by the mobile, except aeronautical mobile, service in region 1 and related studies. In *Provisional Final Acts—World Radiocommunication Conference (WRC-12).* ITU.

ITU-R. (2015). ITU-R Resolution 11–5: Further development of the spectrum management system for developing countries. In *ITU-R Resolutions.* Geneva.

ITU-R. (2019). *Question SG05.77: Consideration of the needs of developing countries in the development and implementation of IMT* Retrieved from www.itu.int

ITU-R. (2020). Appendix 30: (REV.WRC-19) Provisions for all services and associated Plans and List for the broadcasting-satellite service in the frequency bands 11.7–12.2 GHz (in Region 3), 11.7–12.5 GHz (in Region 1) and 12.2–12.7 GHz (in Region 2) In *Radio Regulations.* ITU.

Jakhu, R. S. (2000). *International regulatory aspects of radio spectrum management.* Paper presented at the Workshop on 3G Reforms: Policy and Regulatory Implications, India.

Kenya, Rwanda, & Tanzania. (2003). *Proposal for the work of the conference.* Paper presented at the World Radiocommunication Conference (WRC-03), Geneva.

Kirkaldy, N. (2011). *Mobile broadband.* Paper presented at the ATU Digital Migration and Spectrum Policy Summit, Nairobi.

Lebanon. (2009). *Recommendation ITU-R M.1457–8.* Paper presented at the ITU WP 5D Fifth Meeting. Retrieved from www.itu.int

Levin, H. J. (1981). Orbit and spectrum resource strategies: Third World demands. *Telecommunications Policy, 5*(2), 102–110.

Lyall, F. (2011). *International communications: The international telecommunication union and the universal postal union.* Ashgate Publishing Ltd.

Lyons, P. (2011b). *Harmonization and the economic impact of digital dividend spectrum in Sub-Saharan Africa.* Paper presented at the ATU Digital Migration and Spectrum Policy Summit, Nairobi.

MacLean, D., Souter, D., Deane, J., & Lilley, S. (2002). *Louder voices: Strengthening developing country participation in international ICT decision-making.* Commonwealth Telecommunications Organisation and Panos London.

McCormick, P. K. (2005). The African Telecommunications Union: A Pan-African Approach to Telecommunications Reform. *Telecommunications Policy, 25.* 529-548

McCormick, P. (2007). Private sector influence in the international telecommunication union. *info, 9*(4), 70–80.

Mexico. (2007). *View on Inclusion of a New Radio Interface into ITU-R M.1457*. Paper presented at the ITU WP 8F Meeting. Retrieved from www.itu.int

Mexico. (2011). *Proposal for the Work of the Conference. Agenda Item 1.2*. Paper presented at the World Radiocommunication Conference (WRC-12), Geneva.

Neto, I. (2004). *Wireless networks for the developing world: The regulation and use of licence-exempt radio bands in Africa*. (Master Thesis). MIT.

O'Leary, T., Puigrefagut, E., & Sami, W. (2006). Overview of the second session (RRC-06) and the main features for broadcasters. *EBU Technical Review* (October).

Oberst, G. (2012). *EU and the results of the WRC-12 (The space perspective)*. Current Issues, University of Luxembourg.

Ofcom. (2012). *UK report of the ITU world radio conference (WRC) 2012*. Retrieved from http://stakeholders.ofcom.org.uk

Office of Technology Assessment. (1982). *Radiofrequency use and management: Impacts from the world administrative radio conference of 1979*. Retrieved August 1, 2021 from https://repository.library.georgetown.edu/handle

Pakistan. (2007). *Views on IEEE IP-OFDMA radio interface inclusion to recommendation ITU-R M.1457*. Paper presented at the ITU WP 8F Meeting. Retrieved from www.itu.int

Pavliouk, A. P. (2000). *Incentive radio license fee calculation model*. Retrieved from www.itu.int

Petrazzini, B. (1995). *The political economy of telecommunications reform in developing countries: Privatisation and liberalisation in comparative perspective*. Praeger.

Posta, I., & Terzi, C. (2009). *Effectiveness of the international telecommunication union regional presence*. Retrieved from https://www.unjiu.org

Radiocommunication Bureau. (2007). Report of the director on the activities of the radiocommunication sector on resolution 951. In *World Radiocommunication Conference (WRC-07)*. Geneva.

Radiocommunications Agency. (1997). World radiocommunication conference 1995. *Radiocommunications Agency Business Review 95/96*.

Rouvinen, P. (2004). Diffusion of digital mobile telephony: Are developing countries different? *Telecommunications Policy, 30*(1), 46–63.

Rutkowski, A. M. (1985). Space WARC: The stake of developing countries. *Space Policy, 1*(3).

Rutkowski, A. (1979). 1979 World administrative radio conference: The ITU in a changing world. *Int'l L., 13*(289), 289–327.

Ryan, P. S. (2005). The future of the ITU and its standard-setting functions in spectrum management. In S. Bolin (Ed.), *Standard Edge: Future Generation*: Sheridan Books.

Ryan, P. S. (2012). The ITU and the internet's titanic moment. *Stanford Technology Law Review, 2012*(8), 1–36.

Savage, J. (1989). *The politics of international telecommunications regulation*. Westview Press.

Sims, M. (2007). WRC-07: The technological and market pressures for flexible spectrum access. *Communications and Strategies, 67*, 13–27.

Standeford, D. (2012a). Mobile broadband tops Arab countries' agenda for WRC-12. *PolicyTracker*. Retrieved from www.policytracker.com

Standeford, D. (2012b). WRC agrees on several items but mobile broadband issues remain unresolved. *PolicyTracker*. Retrieved from www.policytracker.com

Standeford, D. (2018). ITU, EU to help African nations clear 700 MHz for 4G. *PolicyTracker*. Retrieved from www.policytracker.com

Sung, L. (2003). Observations from WRC-03. *International Journal of Communications Law and Policy*(8). 1–13

Sung, L. (1992). WARC-92: Setting the agenda for the future. *Telecommunications Policy, 16*(8), 624–634.

UAE. (2009). *View on revision of the recommendation ITU-R M.1457–9*. Paper presented at the ITU WP 5D Fifth Meeting. Retrieved from www.itu.int

UAE, Qatar, & Kuwait. (2011). *Contribution regarding the WRC-12 activities: Agenda item 1.17*. Paper presented at the ASMG 15th Meeting. Retrieved from www.itu.int

United States Department of State. (2003). *United States Delegation Report: World Radiocommunication Conference 2003*. Retrieved from www.ntia.doc.gov

Wang, J. (2018). *Introduction to BSS & FSS plans*. Paper presented at the ITU World Radiocommunication Seminar, Geneva. Retrieved from www.itu.int

Wellenius, B., & Neto, I. (2005). *The radio spectrum: Opportunities and challenges for the developing world*. Retrieved from https://openknowledge.worldbank.org

Wellenius, B., & Neto, I. (2007). *Managing the radio spectrum framework for reform in developing countries*. GICT Publications.

Zacher, M. W. (1996). *Governing global networks: International regimes for transportation and communications*. Cambridge University Press.

Zimbabwe. (2009). *Recommendation ITU-R M.1457*. Paper presented at the ITU WP 5D Meeting. www.itu.int

Chapter 9
The International Spectrum Management Regime

Science should dictate the efficient allocation of spectrum, not politics or international protectionism.
Michael O'Rielly, FCC Commissioner (2016)

9.1 Introduction

The international radio spectrum management regime is one of the oldest still in operation, with the main treaty of the regime, the radio regulations (RR), turning 115 in 2021. However, there has been criticism that the radio sector of the ITU has become increasingly irrelevant to today's wireless world, especially as some countries began to plan their 5G frequencies without waiting for related WRC-19 decisions. But, unlike other international organizations, such as the Security Council, it has been suggested that developing countries feel more empowered in the ITU-R due to the one vote per country rule. Moreover, the presence of hegemonic countries does not necessarily produce an imposed decision—decisions are based on consensus, the objections of a few countries could eventually lead to a compromise being reached that is not wholly aligned with the interests of hegemonic countries. Having said this, while developed countries cannot, in theory, force their decisions on others, they do have the power to lobby and contribute so that their views included in whatever decision is agreed.

In terms of regime theory, there are three main schools of international regime analysis: liberalism (interest-based), realism (power-based) and constructivism (knowledge-based). While (neo)liberalism focuses on the regime's functions, (neo)realism is concerned with the influence of a hegemonic power. In contrast, constructivism analyses regime in terms of cognitive frameworks that influence how actors define problems and their solutions. Neoliberalism has been the dominant within the literature with respect to the telecommunication industry (Ratto-Nielsen, 2006), while portraying the ITU-R as a technical organization that is based on mutual interest.

M. A. El-Moghazi and J. Whalley, *The International Radio Regulations*,
https://doi.org/10.1007/978-3-030-88571-7_9

This chapter, which draws on empirical data from El-Moghazi and Whalley (2019), explores which dominant regime theory applies to modern international spectrum management. It starts with an overview on the different schools of regime theory, and then it examines the international telecommunications and spectrum management regimes, respectively. Following that, it explores the decision-making procedures of the regime and considers examples of changes within the regime.

9.2 Regime Theory

Regimes can be defined as sets of implicit or explicit principles, norms, rules and decision-making procedures around which the expectations of actors converge in a given area of international relations (Krasner, 1982; Zacher, 1996). Zacher (1996) explains that there are hierarchical characteristics associated with principles, norms, rules and decision-making procedures. While principles are general standards of behaviour, norms are the most general prescriptions and proscriptions relevant to an issue and they are implemented at a lower level by rules and decision-making procedures. International regimes are defined as regimes pertaining to activities of interest to members of the international system (Young, 1982). Actors within regimes have an incentive to look for solutions to collective problems (Cowhey, 1990).

There are three main theories when it comes to studying international regimes. Liberalism is based on cooperation and the role of non-governmental actors (Ratto-Nielsen, 2006). Neoliberalism argues that interdependencies and the mutual interests of creating international technical standards could motivate increased international cooperation (McCormick, 2007). In such cases, the gains from cooperation grow sufficiently for countries to be increasingly willing to trade-off policy autonomy for economic welfare (Zacher, 1996).

On the other hand, realism focuses on a country's role and power as the basis of international systems (Ratto-Nielsen, 2006). Neorealists argue that mutual interests are not the crucial foundation on which international regimes are built. Instead, they are built on the dominant country's gains and their power to impose acceptance of the regime and compliance with it on other states (Zacher, 1996). Neorealism also argues that countries, as the main actors within the regime, have as their goal their survival and autonomy and that the distribution of power between states is a major influence on the international regime. It addresses 'structural power', which is power over formal and informal institutional structures (Ratto-Nielsen, 2006).

Constructivism seeks to demonstrate how many core aspects of international relations are given their form by the ongoing processes of social practice and interaction (Bledsoe, 2012). Wendt (1999) explains that there are two main features of constructivism, writing '*that the structures of human association are determined primarily by shared ideas rather than material forces, and that the identities and interests of purposive actors are constructed by these shared ideas rather than given by nature*'. Furthermore, as explained by Puchala and Hopkins (1983) '*regimes accommodate a set of elites who are he practical actors within it as usually, government of Member*

States are the official member of international regimes'. In our case, this is either of ministry responsible for telecommunications, the national regulator, or, in some cases, a specialized agency for spectrum management (e.g. AFNR in France).

While neoliberalism and neorealism embrace different views with regard to regime analysis, both of them adopt a positivist paradigm, which is the dominant paradigm in the field (Lee, 1996). More specifically, they assume that the telecommunications sector is shaped by scientific logic and that the ITU is a technical body. Neoliberalism assumes that telecommunications regulation is employed to minimize barriers to competitive markets, while neorealism argues that the country is the main unit of analysis and that countries compete to maximize their own power. This is not the case for the constructivism approach, which is influenced by the interpretivism paradigm. Under such a paradigm, it is argued that the behaviour of a country is shaped by elite's beliefs and shared knowledge (Ratto-Nielsen, 2006).

In general, liberalism has been the dominant paradigm in the telecommunications sector since 1945. It has been argued that most international regimes are grounded in the mutual interests of countries and not shaped by the strength of the most powerful countries for several reasons. Countries respect the sovereignty of other countries and comply with international agreements. Moreover, there are several international regimes governing different issues that are shaped by the mutual interests of the powerful and weaker countries (Zacher, 1996). While this perspective may explain how the current regime was created, it does not consider changes in (domestic) preferences (Ratto-Nielsen, 2006). Neorealists have explained the change in the international telecommunications regime by reference to the gradual withdrawal of USA from international interconnection (Ratto-Nielsen, 2006). In particular, until the mid-1980s, the ITU recommendations promoted monopolies by allowing telecommunications companies to restrict resale and by prohibiting the interconnection of private networks (Cowhey & Aronson, 1991).

Ratto-Nielsen (2006) criticizes the use of (neo)realism when studying the international telecommunication regime, stating that '*The pessimism of the neorealist view can be attributed to the assumption that states are more concerned with relative gains than absolute gains since international relations are considered as a zero sum game where larger gains from other states increase their advantage in future disputes. Evidently, this has not been the case for the creation and maintenance of the ITU*'.

In contrast, the liberal and neoliberal approaches to the international telecommunications regime view the ITU as a model of international cooperation based on mutual interests (Ratto-Nielsen, 2006). Both approaches, however, have been criticized for failing to take into account the emergence of new policy actors such as NGO or to explain changes within the regime (Ratto-Nielsen, 2006). To this end, constructivism emerged to explain changes in the interests of countries and their preferences that trigger, in turn, changes in the international regime (Ratto-Nielsen, 2006).

Another important perspective is the post-positivism theory of international organizations, which is able, according to some, to overcome the inadequacies of realism and liberalism (Lee, 1996). The proponents of such a theory argue that the ITU has existed as a product of hegemony, engendering rules that facilitated its expansion. Regarding the decision-making procedures, post-positivists argue that the ITU consensus can be challenged by the existence of barely concealed conflicts between Member States. As explained by Lee (1996), who states that '*[i]ndeed, consent to the existing order has been maintained by the ability to issue reservations, serving as a safety valve to diffuse conflict and channel it to statements that do not challenge the existing order*'. It is also perceived that the participation of the private sector reaffirms the hegemony of some actors within the ITU. This is supported by Krasner (1983) who argued that '*hegemonic distribution of power leads to stable open economic regimes because it is in the interest of hegemonic state to pursue such a policy and because the hegemon has the resources to provide the collective goods needed to make such a system function effectively*'.

9.3 International Telecommunication Regime

While the focus of this chapter is on the international spectrum management regime, it is useful to shed some light on the international telecommunications regime that accommodates other functions and stakeholders related to areas such as fibre networks and submarine cables. This will help further our understanding of the ITU.

Until the mid-1980s, the international telecommunications regime focused on three main areas: increasing the reliability of the performance of public service, taking advantage of economies of scale or scope in the provision of services, and promoting the principle of universal service (Cowhey & Aronson, 1991). The main principles of the international telecommunications regime were the free moment of commerce and information, national sovereignty, and one nation not negatively impacting on another (Zacher, 1996). The main norms for the regime were the joint provision of services through the investment by several countries in a common infrastructure, with standardized networks and equipment, and an organized global commons (e.g. spectrum). The regime's decision-making procedures are based on having one vote for each country within the ITU, with three key sets of rules: ITU Convention, Regulations and Recommendations (Cowhey, 1990).

In terms of the main institution for the telecommunication regime, the history of international organizations responsible of telecommunication networks could be perceived through multiple successive periods (MacLean, 2003). In the first period, there were several international organizations responsible of regulating telephone, telegraph and radiocommunication technologies. These were consolidated in the second period into the ITU, with the third period being associated with issues such as privatization, competition and the emergence of new stakeholders.

Until the mid-1980s, ITU recommendations promoted monopolies by allowing telecommunications companies to restrict resale and by prohibiting interconnection of private networks (Cowhey & Aronson, 1991). Ratto-Nielsen (2006) explains that there were several motives behind promoting the old international telecommunications regimes, which was based on the concept of natural monopoly. Not only did governments use the monopoly to build universal telecommunication networks through expanding the network at the lowest cost, but telecommunications networks were also perceived as a key resource, especially in periods of domestic crisis or war, with governments therefore wanting to maintain control of them. Finally, government control was used to support the development and maintenance of national identity.

During the 1980s, the natural monopoly paradigm was challenged by the confluence of technological change (e.g. the integration of computers and telecommunications), changing economic conditions (e.g. public debt), and globalizations (e.g. satellite communications) (Ratto-Nielsen, 2006). Technological innovation also altered the distribution of gains between governments, international business and technology-related industries. New technologies enabled larger firms to deploy their own private internal communications network, which led to the decline of the natural monopoly principle. Furthermore, competition emerged between equipment vendors while previously nationally focused operators began to compete in each other's markets using new technologies like satellite (Ratto-Nielsen, 2006). Around the same time, countries such as the USA, UK and Japan, which had significant international power to stimulate global reform in telecommunications, began to reform their own markets. Moreover, even developing countries were interested in policy reform as part of their strategy to attract foreign investments (Cowhey & Aronson, 1991).

The first challenge to the regime emerged when the FCC extended unilaterally resale and sharing from domestic to international circuits in 1980 (Ratto-Nielsen, 2006). This decision, which was supported by the USA, sought to promote competition in international telecommunications. The liberalization movement spread to other countries so that, in 1988, the ITU World Administrative Telegraph and Telephone Conference (WATTC-88) decided to allow operators and users of private networks the ability to extend their operations internationally. In 1994, telecommunications was included in the General Agreement on Trade in Services (GATS), and in the WTO in 1997 (Ratto-Nielsen, 2006).

9.4 International Spectrum Management Regime

Some have argued that the establishment of the international spectrum management regime at the beginning of the twentieth century was mainly intended to resolve the two issues of interference and interconnection (Anker & Lemstra, 2011). The first radio conference was held in Berlin in 1903 to prevent the Marconi Company from achieving a monopoly over radiotelegraphy, and just a handful of years later, in 1906, another conference in Germany was convened to discuss the same issue along with that of spectrum allocation.

ITU-R is the administrative body responsible for setting the international spectrum management regime's rules, and whose origin can be traced back to the Washington Radio Conference of 1927. This conference established the 'International Radio Consultative Committee' (CCIR) to undertake technical studies in the period between radio conferences. The first time that spectrum was allocated in response to a need rather than legalizing existing uses was in 1938 at the Cairo Radio Conference. The Cairo Conference also established technical standards and restricted the use of radio transmitters (Codding, 1991).

International radiocommunication arrangements can be divided into three main categories (Rutkowski, 1983). The first one is the ITU Convention and constitution, which accommodate principles that address general issues such as the relationship between Member States. According to Article 1 of the ITU Constitution, one of the ITU's purposes is to allocate bands of radio-frequency spectrum, allot radio frequencies and register radio-frequency assignments in order to avoid harmful interference between the radio stations of different countries (ITU, 2011). One other important role can be found in Article 45, which states that all radio stations must be established and operated in such a manner as not to cause harmful interference to other stations that operate in accordance with the provisions of the RR (ITU, 2011).

The RR, which organize the rules of radio operations, is the second category. The RR accommodate the international Table of Frequency Allocations in addition technical, procedural, and operating rules for spectrum (Gregg, 2009). The RR have international treaty status and are, therefore, binding on all ITU-R Member States (Maitra, 2004). The third category is ITU-R recommendations, which are provisions that have no obligatory status.

The regime was based on the concept of the use of common frequencies (Levin, 1971), which can be interpreted as the global harmonization of spectrum allocation. This allocation could be to one or more radio services with equal or different rights (primary and secondary). Secondary services cannot cause harmful interference to primary services and cannot claim protection from harmful interference caused by a primary service (ITU-R, 2008a).

WRC are held every three or four years to revise the RR, which are the international treaty governing the use of radio-frequency spectrum and the geostationary-satellite and non-geostationary-satellite orbits. The conference also addresses any other radiocommunication issue of a worldwide character. The main issues that the WRC studies are organized in terms of what is called an 'Agenda Item' (A.I.). According to Article 7 of the ITU Convention '*[t]he general scope of this agenda should be established four to six years in advance, and the final agenda shall be established by the Council preferably two years before the conference, with the concurrence of a majority of the Member States*' (ITU, 2011).

Consensus on agenda items is usually difficult to reach because excluding an issue from discussion in the forthcoming WRC may lead to the delay of a service or technology, while considering new issues may bring about the threat of harmful interference with existing services (Abernathy, 2004). During the WRC, negotiations are conducted between Member States, which effectively involves them trading support on different issues between one another (Manner, 2003). Decisions on agenda items are usually reached by consensus but, if not, voting occurs where each Member State is entitled to one vote. In addition, each country has the right to make declarations and reservations at the end of the WRC regarding any decision that has been made (Gregg, 2009).

Observers that are sector members of the ITU-R, from the private sector, can attend WRCs in a non-voting capacity (ITU, 2011). Regional organizations usually present common proposals to the WRC on behalf of their members as proposals must have the support of more than one country to be considered (Contant & Warren, 2003). In general, those countries that do not share the view of the others regarding a WRC decision are expected to abide by the opinion of the majority. Those Member States who are unhappy with a decision can record their reservations in the final act of the conference (Jakhu, 2000). The resolution of an issue can occur during the final moments of the WRC, but if this is not possible the decision is usually postponed until the next conference (Gregg, 2009).

It is important to highlight the increasing role of the private sector in the international spectrum management regime, especially since ITU Plenipotentiary Conference 1995 formally recognized the rights of the private sector (MacLean, 1995). The private sector participates in, as well as lobbies, the ITU-R to obtain support for their particular interests (Irion, 2009). Furthermore, although national regulators are the decision-makers for treaty instruments, their decisions are developed based on the work of ITU-R study groups where the private sector conducts much of the technical work (McCormick, 2007). The participation of the private sector is not new. The participation of the private sector can be traced back to the International Telegraph Conference of 1872 that allowed those private companies laying telegraph cables to attend as observers (Ard-Paru, 2013).

The main principles of the international spectrum management regime are twofold. Firstly, the sovereign right of each country to assign its frequencies to any service or station (Lyall, 2011), and, secondly, that all radio stations must not cause harmful interference to other stations which operate in accordance with the provisions of the ITU radio regulations (ITU-R, 2008b). The main norm of the regime is the global harmonization of spectrum allocation. The ITU-R is the administrative cooperation body responsible for setting the regime's rules through the RR, ITU-R resolutions, recommendations and reports. The RR have international treaty status and are binding on all ITU-R countries (Maitra, 2004).

The regime's rules include the registering of national frequencies in the ITU-R MIFR, which is necessary in order to acquire international recognition and thus protection against harmful interference. Such registration requires conformity with the ITU-R service allocation table and not causing harmful interference to existing assignments in other countries (Lyall, 2011). A second rule is the need to allocate each spectrum band to one or more radio services with equal or different rights (primary and secondary). This is based on the results of compatibility and sharing studies that are usually technology-dependent (Louis, 2011). Stations of a secondary service cannot cause harmful interference to stations of primary services and cannot claim protection from harmful interference caused by stations of a primary service (ITU-R, 2008a). A third rule is that of regional allocation, where the ITU divides the world into three regions in terms of spectrum allocation.

Examining the ITU from the perspective of different schools of regime theory reveals two distinct views. From the realist perspective, the ITU was controlled, prior to 1950, by a small number of developed countries (Rutkowski, 1979). These countries used the votes of their overseas colonies to dominate the activities of the ITU. For instance, in 1925, France, Great Britain, Italy and Portugal each controlled seven votes at the ITU (Noam, 1989). In addition, Wallenstein (1977) argues that spectrum allocation is an intensely political exercise of power where worldwide agreement is not always fully achievable.

Adopting a liberalist stance, the creation of the regime was associated with the mutual interest of the founding countries in enabling international interconnection (Codding, 1991). Although the RR are a matter of international law, there is no record of any case of spectrum management compliance coming before an international court (Ryan, 2005). It is also argued that starting from WARC-92, the first international communications conference under the so-called New World Order, tensions between the Soviet bloc and Western nations no longer existed with the consequence that focus was on international cooperation (Sung, 1992). This was due to the reform of the ITU, which was started at the Nice Plenipotentiary Conference 1989 and completed at the Geneva Plenipotentiary Conference 1992, which ended an era of dominance by a few countries (Kelly, 2002). While the technical discussions within the ITU sometimes have a political nature, this has not led to ITU activities being dominated by a handful of states (Savage, 1989).

9.5 Regime Theory Analysis

El-Moghazi and Whalley (2019) undertook an analysis of the international spectrum management regime that identified a number of different views. The first view is that the regime is mainly based on mutual interest due to the need to have common frequencies globally for wireless systems and services such as maritime, aeronautical, mobile and satellite services. In such cases, the interest is considered the self-interest for each country but it is also international in nature as well. Therefore, every country is interested in reaching a common solution. Such a solution is quite often needed in the case of the international spectrum management regime where protection against interference and reaching a certain level of harmonization, whether it is regionally or globally, is in the interest of all countries.

Regarding the accusation that there are dominant countries within the regime, one view is that it depends on the expertise that a country has independent of the country's financial resources or political power. During the history of the ITU-R, there have been several experts from countries such as Morocco and Syria whose expertise and seniority were acknowledged by other countries. Even among leading countries that experience significant political disagreements (e.g. USA and Russia), within the ITU-R there is a need to achieve agreement. In addition, the debates during WRC-12 surrounding 700 MHz and the ability of developing countries to have their views accepted by others showed the benefits that can be achieved through cooperation over countries considered to be dominant or powerful. Furthermore, there has been a transition, with those countries that were previously dominant realizing that they cannot achieve their objectives by forcing decisions on (other) less powerful countries.

The second view supports the theory of dominant countries where dominance is related to the nature of the telecommunications industry and enables some countries to be dominant through developing wireless technologies. Ultimately, the international spectrum regime is related to the telecommunications industry, but it is largely a technical field. Moreover, some countries are dominant by virtue of their economic development, political power and strong telecommunications industry. If the regime had been formalized to decrease the power of these countries, this would deter innovation in these countries which, in turn, may negatively impact on the telecommunications sector as a whole. Dominance is also related to lobbying within the ITU-R, where manufacturing companies and industrial countries usually possess the capabilities to create a strong lobby that supports their interests.

Another observation considers that dominance is not related to the GDP of the country or to its political power. All of the stakeholders are invited to participate in ITU-R decision-making procedures, and those who can attend the meetings, present contributions, participate in the discussions, play a big role in formulizing final decisions. Accordingly, there are number of countries that are considered dominant and influential within the ITU-R (e.g. France, UAE, UK, USA) irrespective of whether they are dominant in other international organizations.

Regarding the third regime theory, constructivism, one view is that while the regime is contribution-driven, the different experts meet and discuss the different proposals in a way that harmonizes the views of participants. Furthermore, some of the individual delegates have the ability to influence their country's positions and even those of other countries. In addition, ideas of a solid nature usually spread among the delegates of different countries and are adopted by them. This is because some country's positions are based on misconceptions or the lack of full information. Therefore, conversations, or sharing specific ideas, can change the position of a country. Individuals can also block the discussion or deter a specific proposal especially if they chair a meeting by delaying giving the floor to participants. Proof of the importance of sharing ideas is evident during WRC, where there is a considerable opportunity for positions to change following various 'elite' meetings.

Additional support for such a view is that the main decisions taken at WRC are through the inter-regional meetings, which are attended by only a few people. Furthermore, some from developing countries are able to defend their interest despite the low participation of these countries and can play an important role in balancing the discussion. These individuals usually familiar with what has happened in past WRCs, which is important when it comes to building arguments.

The fourth view is that the international spectrum management regime displays, at one time or another, features of the three regime theories. Firstly, there are some countries with more resources than others that have more power and influence over discussions. There were instances where these dominant countries forced a decision or introduced an item at WRC without it being on the agenda of the conference. In addition, some discussions are dominated by a few countries due to the nature of these discussions. For instance, only a few countries are active in the field of data satellite or space research, which inevitably limits the number of contributors.

Secondly, there needs to be mutual interest rather than national self-interest if a common understanding or agreement is to be reached regarding a global standard or harmonized allocation. This common ground is usually reached through interactions and sharing ideas between those representing their countries. More specifically, reaching consensus or compromise entails that each country gives up something. This is achieved though discussion and the sharing of ideas.

The international spectrum management regime is also a mix of countries attempting to achieve mutual cooperation, and dominant countries forcing their views and interest on others. For instance, while the system looks like it aims to achieve mutual interest, the collective way of discussing issues at WRC limits the capability of some countries to participate in the different discussions. Additionally, dominant countries attempt to avoid voting as much as possible to show that they are seeking to achieve the benefits for all countries.

This is also the reason why it is rare in the ITU-R to discard one country's proposals or concerns. Countries that have the resources and capabilities usually work to facilitate decision-making process because it is believed this will achieve their objectives more readily. This view supports the existence of a post-positivist stance in regime theory regarding the international spectrum management regime, where hegemonic countries have a mutual interest in a global harmonized spectrum for their technologies, standards and systems. Accordingly, these countries utilize their lobbying power to promote their ideas. They tend to avoid voting because it may negatively influence the regime's strength, which reflects the extent to which ITU Member States abide by the rules.

9.6 Decision-Making Procedures

One of the main elements of the international spectrum management regime is the decision-making procedure where each Member State has one vote. The concept of one vote per country was established by the International Telegraph Conference of 1865 (Ard-Paru, 2013). Countries usually reach decisions without the need for conducting a formal vote, thereby recognizing the importance of consensus in, say, reducing manufacturing costs (Frieden, 2008). However, while voting procedures are well-defined by the ITU-R, the concept of consensus is not formally defined. In particular, Article 32 of the ITU Convention states '*[a]s a general rule, any delegation whose views are not shared by the remaining delegations shall endeavour, as far as possible, to conform to the opinion of the majority*'.

Having said this, according to Article 21 of the ITU Convention, the voting procedure contains details such as expressing reasons for votes by the Member States and the rule that no delegation may interrupt the process once a vote has begun unless it is to raise a point of order regarding how the vote is being conducted. The voting procedure, according to Article 32 of the ITU's convention, also accommodates the possibility of Member States expressing reservations at the end of the conference. Another unique characteristic of the international spectrum management regime is that WRC can partially, or, in exceptional cases, completely revise the RR.

In general, there are two contrasting views regarding the efficiency of decision-making procedures. The first is that these procedures achieve a balance between the needs of industry to innovate and that of governments to regulate. Reaching consensus and agreement between 193 countries is not an easy task, and one that could conceivably take several years. Furthermore, the current character of the procedures creates a lot of certainty and it is necessary to achieve a balance between stability and speed. Secondly, it has also been highlighted that the procedures are one of the most effective and transparent among the UN organizations. Not only are the procedures stated clearly but the topics to be discussed are known to all stakeholders.

Regarding voting, unlike other organizations where decisions can be immediately made subject to voting, the ITU-R allows several other opportunities to reach a decision with the consequence that voting is the last resort. Furthermore, decisions are taken in a 'bottom-up' approach where all the stakeholders participate in the study groups and working parties. Discussions then continue during the WRC until the issue at hand is approved by all Member States. Therefore, as everyone is involved from the beginning, the decision is accepted and implemented globally. Transparency is another advantage of the procedures, with meeting dates and places being known in advance and documents made available to all stakeholders. Equity is also a unique feature of the procedures, with each country having one vote.

The delay in the process is not related to the procedures but to the need to reach a consensus, between neighbouring countries and regional groups. The slow pace of the procedures also reduces, but does not eliminate, the risk of reaching the wrong decision. And with such a slow pace, the procedures are able to accommodate any type of technology development over many years and do not deter their preliminary implementation or development. The decision-making procedures are characterized by their regularity and enable any country to raise an issue and bring it to a study group.

An alternative view criticizes the decision-making procedures. Although decisions related to spectrum allocations are taken at WRCs, these decisions are formulated by study groups where some countries do not participate. Another deficiency is related to the slow pace of the procedures where, in order to introduce a new agenda item to WRC, a period of eight years is needed. Even if there is an urgent topic that requires holding a conference between two WRCs, it will interrupt the study period and may not be possible due to the need for extensive logistical preparations to be made. In the period, before each WRC the issues under study may undergo changes that distance it from the original issue. Moreover, one country can block another for political reasons as the procedures enable them to do so through, for example, introducing a country footnote that requires the approval of neighbouring countries.

It seems that countries with access to more resources, in terms of experts, participation and contributions, can benefit the most from the decision-making procedures. Therefore, developed countries, as well as significant players in the telecommunications industry, are the main beneficiaries of the procedures. A critical deficiency of the procedures is that non-compulsory documents, which may be important and influential, are approved during the study groups meetings rather than during the WRC and not every country attends and follows their development.

In conclusion, it seems that the views on the regime decision-making procedures depend on where you stand. Industry and developed countries call for a quicker pace of WRCs in order to have their technologies adopted by the ITU-R and deployed around the globe. On the other hand, developing countries, who are typically not leading the discussion, need more time to replace their old equipment. Regarding the procedures themselves, while they seem fair and inclusive, they also are strongly influenced by discussions at the study groups, which are held prior to WRCs, typically in Geneva (Switzerland). This may significantly deter the involvement of those countries, that lack the necessary financial resources and technical expertise in the decision-making process. Another difficulty is related to the way issues are discussed during WRCs, where quite often more than 20 agenda items are discussed simultaneously. This allows one country to block another for political reasons and also to trade positions on agenda items. It also limits the efficient participation to those countries that have a large number of delegates.

9.7 Regime Change

Several incidents in the last decade call for examining whether the international spectrum management regime is changing and questioning whether the ITU-R is not as important as it previously was. One of the critical incidents was the threat by the FCC chairman, Michael O'Rielly, in 2016, following WRC-15, that the USA may consider withdrawing its funding from the ITU. In his own words: '*I will not hesitate to advance the United States' technological positions to ensure future successes—with or without the ITU ... global technological leaders, such as the U.S., will continue to innovate outside and without input from the ITU and its many nation states. This will, in turn, make the ITU and the WRC process less relevant*' (O'Rielly, 2016).

Another incident is related to 5G frequency allocation. European countries decided to use the 26 GHz frequency band for 5G, albeit with certain conditions, without waiting for WRC-19 to decide whether the band would be identified for IMT under Agenda Item 1.13. In contrast, the USA, Japan and South Korea, following their failure to include the 28 GHz band as one of the potential bands to be identified for IMT in WRC-15, decided to utilize the band for 5G even if the band is not considered or approved.

These incidents force us to reconsider the positions of the main stakeholders in the international spectrum management regime, and whether there have been changes to its main norms and principles. The USA is one of the main contributors of the ITU, and it is quite common for the USA to have the largest number of delegates with an ambassador leading them at WRC. In contrast, the European countries were the ones who created the regime and established the ITU-R, which has been considered by some in the past to be European organization. There is, therefore, a need to understand how the main stakeholders perceive the regime and to ascertain if it has changed.

There appears to have been changes in the power balance within the ITU-R, at least when it comes to mobile telecommunication services. This was evident in WRC-12, when the Arab and African countries managed to force the European countries to accept the 700 MHz mobile allocation without it being an agenda item of the conference. It was also evident when significant contributors to the ITU, such as the USA and Japan, failed to include the 28 GHz band as one of the candidate bands for IMT in WRC-15. Two questions arise: What do these incidents indicate, and why are these countries eager to participate in ITU-R activities?

Firstly, regional groups have started to play a more substantial and influential role in the ITU-R. Such a role is, however, largely dependent on the number of votes these regional groups have. The African and Arab regional groups have realized that they have growing power within the organization. Historically important contributors to the ITU, like Japan and the USA, may face a more difficult situation within their regional groups in future as the developing countries within APT and CITEL, respectively, have also started to be active and contribute more to discussions. In addition, while the Arab and African regional groups appear to be more homogenous in terms of GDP and are an importer of technologies; this is not the case for other regional groups, which makes coordination for countries such as the US more difficult.

So why are developed and developing countries keen to participate in the ITU-R? Each of the stakeholders has a different reason of being part of the ITU-R. Firstly, with respect to developing countries, the ITU-R is an important place for them where they feel empowered by one vote for one country, and they act on an equal basis with developed countries. It is not, therefore, a surprise that the ITU has the greatest developing country involvement in terms of their numbers compared to other international organizations (MacLean et al., 2002). Another important factor for developing countries is that they usually only accept standards that have been approved by the ITU-R (e.g. IMT). Otherwise, they risk investing in a technology that could turn out to be a failure.

With respect to developed countries, if we consider them as the dominant ones within the ITU-R, why they are also keen to contribute financially to the organization and be part of it? Why, for example, did the USA not withdraw from the ITU-R following WRC-15 in accordance with the threats made by the FCC chairman when the US left UNESCO? Developed countries are quite keen to be part of the ITU process for several reasons. These industrial countries want to push their standards to gain a certification that it has the approval of more than 190 countries. Secondly, as the ITU-R does not have stringent enforcements measures in cases of interference and jamming, it is in the interests of developed countries to comply with a regime whose rules are widely accepted. This would be achieved by showing how important such compliance is, and how following the RR is part of country's wider obligations. It could also be argued that developed countries want the developing countries to feel that they are included within the regime. This is why it is very common in the ITU-R to find countries participating in discussions although that is not related to them. In other words, some participating countries act to demonstrate that they are part of the regime. One other reason for cooperation between the different countries in the ITU-R is related to the nature of radio spectrum, which imposes cooperation and coordination on countries. This was, of course, the reason why the regime emerged in the first place.

Another area for examination is whether the regime is experiencing changes and if these alter the regime's main principles and rules. One must admit that there is an increasing trend of countries not waiting for the ITU-R, as illustrated by developments such as the 26 GHz band mobile identification in CEPT before WRC-19 or the case of USA and Japan operating IMT in bands not identified for it by the ITU-R (e.g. 28 GHz). Meanwhile, it appears that forums such as the GSMA annual exhibition in Barcelona and 3GPP meetings have become more popular in the telecommunication sector at the expense of CPM and WRC meetings. Having said that, the international spectrum management regime is still relevant to the main stakeholders and it is likely that it will survive without significant alterations to its main principles or norms. Even if some countries decide to act outside of the IMT identification or not to wait for WRC decisions, this is mostly conducted within the limits of mobile service allocation. In other words, using cellular mobile in a band allocated to the mobile service but not identified to IMT is possible according to the RR, which indicates that IMT identification does not preclude other uses for the band.

Secondly, the telecommunications industry still needs the RR as it is an international treaty, which is respected by more than 190 countries and decides what uses radio spectrum will be put to. Thus, even if countries are sovereign and can act outside of the RR, they prefer not to do so. As explained by Ryan (2005), although the RR are a matter of international law, there has never been a single case of a spectrum management dispute appearing in an international court. The durability of the regime also indicates the common benefits shared by all ITU-R Member States. As outlined by Zacher (1996), there are characteristics that indicate the existence of mutual interests:

- The durability of a regime and its norms over a long period of time during which there were significant changes in the international regime.
- Adherence to the regime by countries that are hostile to each other elsewhere.
- Most or all countries gain from the regime and no evidence that one group of countries lose a great deal as a result of the regime and
- There is no indication that one group of countries uses sanctions to force another group to join the regime or to comply with it.

Examining the international spectrum regime shows that all these conditions are met. The regime is one of the oldest managing a global common resource and survived two world wars and the collapse of the Soviet Union. Secondly, as was argued earlier in this chapter, political conflict does not lead to non-cooperation within the ITU-R. It is very common during WRC to find conversation and coordination between countries that have political disagreements between them. Further, no regional group or a country seems able to capture the bulk of the benefits that accrue from the regime. Finally, complying with the RR is not enforced, but yet accepted and followed by the majority of countries.

But can the international spectrum regime be altered or changed? Generally speaking, changes to international regimes occur when there is a new coalition that alters domestic regulatory bargains in countries with significant influence on the world stage. Those countries are unlikely to accept international regimes that collide with their domestic preferences (Cowhey, 1990). To date, this is not the case for the international spectrum regime, as, domestically, spectrum is still mostly managed according to the traditional command and control regime except for measures such as an auction or technology neutrality. These measures do not contradict with the principles and norms of the international regime.

A second cause that may lead to significant alterations in the regime's structure of rights and rules are shifting in the underlying structure of power that makes it more difficult to impose decisions as the dominant actor's power declines (Young, 1982). While the balance of power is not as it was at the time of the regime's emergence, when European countries were dominant, it is now increasingly exercised through regional groups exploiting the one vote per country rule. Such a rule represents a fundamental principle in international relations, namely a country's national sovereignty where the state is the main actor in UN organizations and cannot be forced to adopt specific regulations.

In fact, such a principle is essential to spectrum management due to the historical desire of national regulators to control access to their radio spectrum as a national resource, as well as the issue of cross-border interference. Drake (2008) also argues that the waves of liberalization and privatization have not been able to alter the international spectrum management regime. However, they have led to the increasing role of the private sector in ITU study groups and its main conferences. And, of course, a country's position in the ITU is also likely to be influenced by their companies.

However, it is necessary to note there have recently been changes among the main stakeholders in the ITU-R. The first change is related to the emergence of China and India as large telecommunication markets reflecting the size of their populations. The second one is also related to the rise of the African and Arab regional groups as block of votes within Region 1, and the rise of other countries (e.g. Brazil) within CITEL which has made regional coordination by the USA more difficult. The third possible cause for change is technological developments (Young, 1982). While technologies such as CRS and SDR were perceived as altering the way countries manage the radio spectrum, so far this has not been the case (at least internationally).

This is all not to deny that the ITU as evolved as the product of several interests and ideas (Lee, 1996). The ITU has been through major changes, reflecting developments within the global political economy in general and the telecommunications industry in particular. This can be seen in the increasing participation of the private sector and empowerment of developing countries. However, focusing on the international spectrum regime, the main norms and principles have remained largely the same since the regime commenced except for cases such as the increasing use of post-priori planning instead of a priori planning in certain service allocations like satellite broadcasting.

In general, this chapter has shown that none of the different schools of theory can solely capture the essences of the international spectrum management regime. This provides empirical support to the need to synthesis across these different schools of thought. In addition, the technology development and inadequacy of the ITU-R decision-making procedures could be perceived differently according to which radio-communication service is under consideration. For instance, while cellular mobile service have rapidly grown and developed, this may not be the case for services such as radiolocation or maritime where the development is somewhat slower and the systems stable. An example of the existence of the liberalism and realism schools within the ITU-R is the rivalry between the different radiocommunication services in the same frequency band, which is based on the relative gain that is supported by realism while harmonization between countries for the same service is based on absolute gain as explained by liberalism.

Thirdly, the policymakers have an important role in discussions, especially in meetings towards the end of WRC where the attendance is limited to a few representatives from each regional group. Meanwhile, sharing ideas between policymakers is the motivation for changes in developing countries where there is a weak institutional system, because these countries usually lack the resources to participate in the meeting or a clear self-interest to determine their goals within WRC.

9.8 Summary

The international radio spectrum management regime is one of the oldest existing regimes with the main treaty of the regime, the RR being 115 years old in 2021. It has, however, been argued that the ITU-R has become irrelevant in today's wireless world (O'Rielly, 2016). While the regime's decision-making procedures have begun not to reflect advances in wireless technologies, most of the telecommunication sector's leading actors prefer to operate according to its procedures. This provides them with ITU-R approval for their technologies, which facilitates the acceptance of their technologies in (developing) countries. Moreover, while the pace of technological change may now be faster than ITU-R procedures, the ITU-R system is still able to accommodate these changes, and even if countries are not satisfied with the performance of the ITU-R, it remains in their interest to include their regulations in the RR. This helps perpetuate the relevance of the ITU.

The ITU also continues to be relevant because of how it is organized. Through one vote per country, developing countries feel empowered. Leading countries may argue and lobby for their positions, but one vote per country necessitates compromise and consensus between Member States. Moreover, the voting procedure has also encouraged and empowered the emergence of regional groups—this reflects the numerical superiority of developing countries within the ITU, but increasingly their divergent spectrum needs and requirements compared to developed countries.

As a consequence, none of the three schools of theory outlined in the chapter are able to solely capture the essence of the international spectrum management regime. While the international spectrum management regime may provide further opportunities for investigation, these need to reflect the significant role that context plays in shaping it. The rivalry between different telecommunication services in the same frequency band may be explained by realism, but it does depend on the exact nature of the services and their spectrum requirements. Similarly, harmonization may be explained by liberalism, but its exact nature depends on the spectrum bands and services in question.

References

Abernathy, K. Q. (2004). Why the world radiocommunication conference continues to be relevant today. *Fedral Communications Law Journal, 56*(2), 287–299.

Anker, P., & Lemstra, W. (2011). Governance of radio spectrum: License exempt devices. In W. Lemstra, V. Hayes, & J. Groenewegen (Eds.), *The innovation journey of wi-fi: The road to global success*. Cambridge University Press.

Ard-Paru, N. (2013). *Implementing spectrum commons: Implications for Thailand*. Ph.D. Thesis. Chalmers University of Technology.

Bledsoe, J. (2012). *Exploring the relationship between uneven economic development, racial and religious intolerance and ethnic cleansing*. Fourth Annual Student Research Conference-Suffolk University Government Department.

Codding, G. A. (1991). Evolution of the ITU. *Telecommunications Policy, 15*(4), 271–285.

Contant, C. M., & Warren, J. (2003). The world radiocommunication conferences process: Help or hindrance to new satellite development? *Acta Astronautica, 53*(4).

Cowhey, P. F. (1990). The international telecommunications regime: The political roots of regimes for high technology. *International Organization, 44*(2), 169–199.

Cowhey, P., & Aronson, J. D. (1991). The ITU in transition. *Telecommunications Policy, 15*(4), 298–310.

Drake, W. J. (2008). Introduction: The distributed architecture of network global governance. In W. J. Drake, & E. J. Wilson (Ed.), *Governing global electronic networks: International perspectives on policy and power*. MIT Press.

El-Moghazi, M., & Whalley, J. (2019). *The international radio spectrum management: A regime theory analysis* 30th European Regional ITS Conference.

Frieden, R. (2008). Balancing equity and efficiency issues in the management of shared global radiocommunication resources. In W. J. Drake & E. J. Wilson (Eds.), *Governing global electronic networks: International perspectives on policy and power.* MIT Press.

Gregg, D. C. (2009). Lessons learned from the spectrum wars: Views on the United States' effort going into and coming out of a world radiocommunication conference. *CommLaw Conspectus: Journal of Communications Law and Policy, 17*(2), 377–415.

Irion, K. (2009). Separated together: The international telecommunications union and civil society. *International Journal of Communications Law and Policy, 13*, 95–113.

ITU. (2011). *Constitution of the international telecommunication union* (Collection of the basic texts of the ITU adopted by the plenipotentiary conference (Ed 2011)). ITU.

ITU-R. (2008a). Article 5: Frequency allocations. In *Radio regulations* (Vol. 1). ITU.

ITU-R. (2008b). *Preamble* In *Radio regulations* (Vol. 1). ITU.

Jakhu, R. S. (2000). *International regulatory aspects of radio spectrum management*. Workshop on 3G reforms: Policy and Regulatory Implications. Centre for the Study of Regulated Industries, McGill University.

Kelly, T. (2002). Never-ending international telecommunication union reform. In R. Mansell, R. Samarajiva, & A. Mahan (Eds.), *Networking knowledge for information societies: Institutions & intervention*. Delft University Press.

Krasner, S. D. (1983). Regimes and the limits of realism: Regimes as autonomous variables. In S. D. Krasner (Ed.), *International regimes (Cornell Studies in Political Economy).* Cornell University Press.
Krasner, S. (1982). Structural causes and regime consequences: Regimes as intervening variables. *International Orginzation, 36*(2), 185–205.
Lee, K. (1996). *Global telecommunications regulation: A political economy perspective.* Pinter.
Levin, H. J. (1971). *The invisible resource.* Johns Hopkins Press.
Louis, J. (2011). *International radio spectrum management beyond service harmonisation.* Fourth international conference on emerging trends in engineering & technology. Port Louis, Mauritius
Lyall, F. (2011). *International communications: The international telecommunication union and the universal postal union.* Ashgate Publishing Ltd.
MacLean, D., Souter, D., Deane, J., & Lilley, S. (2002). *Louder voices: Strengthening developing country participation in international ICT decision-making.* Commonwealth Telecommunications Organisation and Panos London.
MacLean, D. (1995). A new departure for the ITU: An inside view of the Kyoto plenipotentiary conference. *Telecommunications Policy, 19*(3), 177–190.
MacLean, D. (2003). The quest for inclusive governance of global ICTs: Lessons from the ITU in the limits of national sovereignty. *Journal of Information Technologies and International Development, 1*(1), 1–18.
Maitra, A. (2004). *Wireless spectrum management. Policies, practices, and conditioning factors.* The McGraw-Hill Companies.
Manner, J. A. (2003). *Spectrum wars: The policy and technology debate.* Artech House.
McCormick, P. (2007). Private sector influence in the international telecommunication union. *info, 9*(4), 70–80.
Noam, E. M. (1989). International telecommunications in transition. In R. W. Crandall & K. Flamm (Eds.), *Changing the rules. Technological change international competition and regulation in communications.* The Brookings Institution.
O'Rielly, M. (2016). 2015 *World radiocommunication conference: A troubling direction FCC.* Retrieved 30 June 2016, from www.fcc.gov
Puchala, D. J., & Hopkins, R. F. (1983). International regimes: Lessons from inductive analysis. In S. D. Krasner (Ed.), *International regimes (Cornell Studies in Political Economy).* Cornell University Press.
Ratto-Nielsen, J. (2006). *The international telecommunications regime: Domestic preferences and regime change.* Retrieved from lulu.com.
Rutkowski, A. (1979). 1979 world administrative radio conference: The ITU in a changing world. *Int'l L, 13*(289), 289–327.
Rutkowski, A. M. (1983). Deformalizing the international radio arrangements. *Telecommunications Policy, 7*(4), 309–316.
Ryan, P. S. (2005). The future of the ITU and its standard-setting functions in spectrum management. In S. Bolin (Ed.), *Standard edge: Future generation.* Sheridan Books.
Savage, J. (1989). *The politics of international telecommunications regulation.* Westview Press.
Sung, L. (1992). WARC-92: Setting the agenda for the future. *Telecommunications Policy, 16*(8), 624–634.
Wallenstein, G. (1977). Development of policy in the ITU. *Telecommunications Policy, 1*(2), 138–152.
Wendt, A. (1999). *Social theory of international politics.* Cambridge University Press.
Young, O. R. (1982). Regime dynamics: The rise and fall of international regimes. *International Organization, 36*(2), 277–297.
Zacher, M. W. (1996). *Governing global networks: International regimes for transportation and communications.* Cambridge University Press.

Chapter 10
World Radiocommunication Conference-19

The type of discussions we have at WRCs are really paving the way for decisions that will be there for 20 or 30 years.
Francois Rancy, ITU-R Director (2015)

10.1 Introduction

This chapter focuses on the most recent World Radiocommunication Conference, WRC-19, which was held in Egypt and lasted for around a month. Not only was this the first WRC since 2000 to be held outside Geneva (Switzerland), but more than 3500 participants attended. It is also arguably one of the most successful, with the conference activities actually ending early on the evening of the last day—this is unlike previous WRC, where activities of the final plenaries lasted until the next morning. Moreover, voting was not used in the WRC, with a consensus being reached across all agenda items.

The broad nature of the topics discussed at WRC-19 is arguably beyond the scope of a single book. This chapter will focus on the different agenda items of WRC-19 that are related to discussions of RR reform such as AI 9.17 that looked at spectrum assignments of satellite earth stations nationally. With this in mind, the chapter begins with an outline of relevant activities from Radio Assembly 2019 (Sect. 10.2) before switching to WRC-19 (Sect. 10.3). This is then followed by a brief overview of two future WRC in Sects. 10.4 and 10.5. The final section offers a summary of the chapter.

10.2 Radio Assembly 2019

While WRCs receive a lot of the attention from the ITU-R community, the Radio Assembly (RA) is an important meeting that is held one week immediately before the conference. The RA is responsible for the structure, programme and approval of

M. A. El-Moghazi and J. Whalley, *The International Radio Regulations*,
https://doi.org/10.1007/978-3-030-88571-7_10

radiocommunication studies and may also approve those ITU-R Recommendations and Questions developed by the Radiocommunication Study Groups (ITU-R, 2015a).

The RA of 2019 was held between 21 and 25 October 2019, in Sharm El-Sheikh City, Egypt. It approved a range of ITU-R Recommendations and Resolutions including those on short-range devices, radio-frequency identification devices (RFID) and railway radiocommunications systems between train and trackside (RSTT) (ITU, 2019b).

One of the issues that was addressed by RA-19 was broadcasting. Although the idea of having a global broadcasting standard has been around for quite a few years, this was the first time it has been written down as an objective. At the moment, there are four main bodies responsible for standardization globally: ATSC (Americas), DVB (Europe), ISDB (Japan) and DTMB (Asia). Having one global standard would achieve economies of scales and facilitate the roaming of TV receivers (Marti, 2019a).

RA-19 approved two resolutions. The first one—'Principles for the future development of Broadcasting'—will develop recommendations and reports facilitating the introduction of new systems, technologies and applications for broadcasting (ITU-R, 2019c). These would help to achieve the global harmonization of specifications, while taking into account the requirements and situations of individual countries and regions.

The second resolution—'Role of the Radiocommunication Sector in the ongoing development of Television, Sound and Multimedia Broadcasting'—sought the development of a roadmap for ITU-R activities for broadcasting by the relevant Radiocommunication Study Group (ITU-R, 2019f). Significantly, the work should be undertaken in conjunction with other ITU-R study groups, ITU-T and ITU-D as well as organizations external to ITU (ITU-R, 2019e).

It is important to highlight that sometimes discussions at a Radio Assembly are related to the following WRC. For instance, administrations may be keen to approve a recommendation or a resolution that would support a certain position at WRC. Having a resolution from the RA could indicate no change at WRC, or, alternatively, a recommendation could motivate WRC to incorporate it by reference so it would have a more obligatory status. On the other hand, countries may delay the adoption of certain recommendation as it may affect what happens at the WRC.

With that in mind, RA-19 discussed the draft ITU-R recommendation developed by Study Group 7. This recommendation was sent for adoption by correspondence but was not adopted due to opposition from the USA (Chairman Radiocommunication Study Group 7, 2019). The recommendation aimed to create rules for IMT and ESSS coordination. During RA-19, some countries sought the approval of this recommendation by the Assembly as it would facilitate the decision that could be taken by WRC-19 with respect to Agenda Item 1.13. But, in contrast, some other countries opposed the approval of the recommendation and asked for more consideration of it by SG 7 in coordination with SG 5. Accordingly, RA-19 decided to instruct SG 7 to continue development of the recommendation during the next study cycle (ITU-R, 2019i).

Another topic that attracted a lot of attention, and perhaps required the largest number of meetings during RA-19, was the adoption of the revised version of ITU-R

Recommendation 1036 that detailed IMT frequency arrangements. Previously, WP 5D was not able to reach a consensus on some points during the study cycle prior to WRC-19, and, accordingly, continued discussion was postponed until RA-19. One of these points was a discussion of the following paragraph, which was below the table stating the footnotes identifying the band for IMT, which was published in the latest published version of the recommendation: '*[a]lso, administrations may deploy IMT systems in bands allocated to the mobile service other than those identified in the RR, and administrations may deploy IMT systems only in some or parts of the bands identified for IMT in the RR*' (ITU-R, 2015f).

The issue was discussed during SG 5 in September 2019 prior to RA-19, and due to the variance in views, SG 5 decided, in accordance with procedures, to forward the draft revision of Recommendation ITU-R M.1036–5 to the Radiocommunication Assembly for consideration (Chairman Study Group5, 2019). During RA-19, after several rounds of discussion, it was agreed to modify the text and to have it as part of an attachment to the recommendation—it would not belong to the *recommends* part but, instead, is considered as a descriptive part of the Recommendation (ITU-R, 2019i).

10.3 World Radiocommunication Conference 2019

WRCs are normally conducted every three or four years and revise the RR and any associated frequency assignment/allotment plans. In addition, according to Article 7 of the ITU Convention, the agenda of a WRC may include '*the partial or, exceptionally, complete revision of the RR, any other question of a worldwide character within the competence of the conference, an item concerning instructions to the Radio Regulations Board and the Radiocommunication Bureau regarding their activities, and a review of those activities; the identification of topics to be studied by the radiocommunication assembly and the radiocommunication study groups, as well as matters that the assembly shall consider in relation to future radiocommunication conferences, and any question which a Plenipotentiary Conference has directed to be placed on the agenda* (ITU, 2019a)'.

According to Article 7, the general scope of the agenda is established four to six years in advance, and the final agenda established by the Council two years before the conference. This means, in practice, that each WRC usually includes a draft agenda for the next two WRCs. For example, WRC-19 decided the agenda of WRC-23 and the preliminary agenda of WRC-27. A degree of flexibility in the agenda of conferences is still possible. WRC-23 is expected to accommodate urgent issues, while WRC-27 will include items that were deferred to future conferences (ITU, 2019a).

The topics discussed during WRC-19 can be categorized into six categories: land mobile and fixed services, broadband applications in the mobile service, satellite services, science services, maritime, aeronautical and amateur services, and general issues. The fact that the ITU is a UN organization indicates its global influence, providing its output with more credibility. In particular, modifications to the RR under WRC-19 agenda items, which may seem purely technical, are in many cases related to the UN Sustainable Developments Goals (SDGs). For instance, SDG 9, which is to increase access to ICTs and strive to provide universal and affordable access to the Internet in least-developed countries by 2020, is related to radio systems such as HAPS and NGSO which increase access to connectivity. EESS, which can monitor natural disasters, is related to SDG 13 (climate change) (Maniewicz, 2019b).

In addition, while the ITU is considered as a technical international body, given that it is still an UN organization, politics are sometimes involved in WRC discussions. For instance, in WRC-15, a revised version of Resolution 12 ('Assistance and support to Palestine') sought to continue assisting Palestine in order for it to obtain and then manage spectrum (ITU-R, 2015b). Palestine and Israel reached an agreement regarding the assignment of radio frequencies for exclusive use by Palestinian cellular operators as well as on a shared basis by Palestinian and Israeli operators (ITU, 2015a).

The Arab countries proposed that the conference should call on the occupying power, as they described Israel, to eliminate all violations of the frequency spectrum belonging to the State of Palestine (ASMG, 2019b). The Arab countries also called for Israel to end the restrictions that it imposed which limited the importation of components (ASMG, 2019b). They also demanded that a technical committee is formed within the Radiocommunication Bureau to investigate the measures taken by Israel and the illegal coverage by its operators of Palestinian territory and assess the losses that have occurred (ASMG, 2019b). Eventually, WRC-19 agreed to review Resolution 12 to enable Palestine to continue developing its 3G and 4G systems while urging all those concerned to facilitate the importation of equipment into Palestine. WRC-19 also required the director of the BR to report to the WRC-23 on what progress has been achieved (ITU-R, 2020c).

It is a usual practice for countries to include some sort of reservation in the final act of WRC. Reservations typically provide countries with the right to act within its national territory to protect itself against any signal that is incompatible with its sovereignty or which may endanger its security or come into conflict with its cultural heritage. Other reservations can address a specific agenda item or topic. For instance, Egypt had a reservation regarding the BR application of Rule of Procedure 3.3c as a result of WRC-15 A.I. 1.6.1. Another interesting reservation was made by several eastern European countries with regard to the L-band, regretting that it was not possible to reach a mutually acceptable agreement regarding the shared use of the 1427–1518 MHz frequency band with their neighbouring countries for aeronautical mobile services.

Sometimes reservations have a political flavour. For instance, Cuba made a reservation at WRC-15 to protect itself against signals which were incompatible with its sovereignty or which may be a danger to its security. The USA, in turn, stated its right to broadcast to Cuba on appropriate frequencies free of jamming and reserved its right to act in future if Cuba interfered with its broadcasting (ITU-R, 2012).

It is important to highlight the role of the RRB during WRC, where the former can submit to the later proposals for modifications to the RR. During the conference, a RRB member can attend the conference in an advisory capacity to provide advice regarding the difficulties associated with the application of regulatory provisions and possible modifications to the RR (Jeanty, 2019). The ITU-R BR director also has an important role to play in the conference, reporting errors, inconsistencies and outdated provisions within the RR (ITU-R BR Director, 2019). The director can also inform the conference about BR's experience of the application of the RR and the difficulties that have been encountered (ITU-R BR Director, 2019).

Another role of the RRB is with regard to the Plenipotentiary Conference (PP) of the ITU. This can instruct the WRC to include a specific matter in its agenda. In fact, that was the case for WRC-15, which was instructed by the PP-14 to include in its agenda global flight tracking (ITU, 2014). This was in response to the loss of Malaysian Flight MH370. Accordingly, WRC-15 allocated radio-frequency spectrum to global flight tracking in the civil aviation station (ITU, 2015b).

One of the permanent items on all WRCs agendas is A.I. 8, which addresses requests from countries to delete their footnotes or to have their name deleted from footnotes if they are no longer required (ITU-R, 2016c). Countries can also add their names to a new or existing footnote subject to several conditions (ITU-R, 2020d). In the following subsections, we will focus on the most important topics of the conference.

10.3.1 International Mobile Telecommunications

Agenda Item 1.13 considered the identification of frequency bands for the future development of IMT, including the possible additional allocation to mobile service on a primary basis in accordance with Resolution 238 (ITU-R, 2016b). This resolution called for studies to determine the spectrum needs for the terrestrial component of IMT in the frequency range between 24.25 GHz and 86 GHz. Sharing and compatibility studies were also required in a large number of frequency bands that are allocated to mobile services on a primary basis. Eventually, the conference identified 17.25 GHz of spectrum, with 14.75 GHz of this harmonized worldwide (ITU, 2019e). While identifying the 24.25–27.5 GHz, 37–43.5 GHz, 45.5–47 GHz, 47.2–48.2 and 66–71 GHz frequency bands for the deployment of 5G networks, WRC-19 also sought to protect EESS.

One disputed issue was IMT identification in the 24.25–27.5 GHz band. In particular, there was a tense discussion with regard to the protection of EESS and its satellite passive sensors which operate in the 23.6–24 GHz band to OOBE in the 24.25–27.5 GHz band. The protection criterion of the EESS is −166 dBW/200 MHz, which may only be exceeded for 0.01% time and 0.01% area (IDate & Plum, 2019). The views of regional groups varied significantly before the conference, with the least restrictive conditions being proposed by CITEL and slightly difference ones being suggested by ASMG and ATU (Kraemer, 2019).

Meanwhile, the World Metrological Organization (WMO) supported the unwanted emission levels of −55 dB (W/200 MHz) for base stations and −51 dB

(W/200 MHz) for user equipment to protect EESS (WMO, 2019). Furthermore, the Space Frequency Coordination Group (SFCG) was concerned that IMT-2020 systems would be unable to comply with the unwanted emission levels identified in the studies. These levels were determined to be in the range of −55 to −49 dBW/200 MHz for base stations, and −51 to −45 dBW/200 MHz for user equipment (SFCG, 2019a).

Prior to WRC-19, the satellite industry was worried about their investments and called for no constraints on the evolution of existing satellite services to be caused by the introduction of IMT (Global Satellite Coalition (GSC), 2019). A study, endorsed by the satellite industry, argued that typically only half of the spectrum harmonized for use in a particular region had been licensed to mobile operators. It did, however, vary significantly from region to region. The study also showed that the amount of spectrum licensed is roughly a third of the ITU's estimation of IMT spectrum demand in 2020 (LS Telecom, 2019).

During the conference, Russia presented the results of a IMT-2020 base station measurement study that questioned whether the interference from actual IMT-2020 networks could be significantly higher than the interference calculated from the reference networks used in various ITU-R studies. Hence, they suggested a limitation on base station unwanted emissions (Russia, 2019). Eventually, it was agreed to adopt a novel two-stage approach and include them in Resolution 750. More precisely, a limit of –33/–29 dBW/200 MHz applies for base/mobile stations brought into use before 1 September 2027, and a limit of –39/–35 dBW/200 MHz for base/mobile stations brought into use after 1 September 2027. The main motivation behind such an approach was that the less stringent unwanted emission limits would be applied when there are fewer IMT deployments, and hence, less aggregate interference (ITU-R, 2019d).

The decisions of WRC-19 regarding the 26 GHz band, and the results which were not originally the proposal of any regional groups or country at the conference, show the importance of globally harmonized conditions for IMT usage. However, the WMO expressed their view that such conditions may lead to ten times more interference from out-of-band emissions than has been recommended and thus threaten, as a result, critical national weather warning systems (Viola, 2020).

Another interesting discussion was regarding the identification of IMT in the 66–71 GHz band. Prior to the conference, the CITEL regional group was not in favour of identifying the 66–71 GHz band for IMT due to the usage, or planned usage of the band, for license-exempt technologies such as multiple gigabit wireless systems (MGWS). CEPT supported equal access to the band by both IMT and MGWS (SFCG, 2019b). The RCC did not support IMT identification due to its incompatibility with existing primary radio services (SFCG, 2019b). In contrast, ASMG supported identifying the band to IMT while highlighting other technologies in a WRC Recommendation (ASMG, 2019a).

During the conference, there was a debate between two groups. The first, led by the USA, called for equal access to the band by MGWS and IMT given that the former is recognized by ITU-R Recommendation M. 2003–2 to potentially operate globally in the 60 GHz band (ITU-R, 2018a). The second, led by the Arab countries, argued that MGWS is not defined within the RR and that they would consider any identification to be outside the scope of the conference. Ultimately, WRC-19 identified the 66–71 GHz band for IMT. The resolution clarified also that countries can also use the band to implement other applications of the mobile service including other wireless access systems (ITU-R, 2019h).

10.3.2 Footnote 5.44B1

One issue that surprisingly drew much attention throughout the conference was IMT identification in the 4800–4990 MHz bands. The origin of the issue was when WRC-15 identified the band for IMT in three Asian countries (Laos, Vietnam and Cambodia) subject to strict conditions to protect aeronautical mobile service (AMS) (ITU-R, 2016a). The 4.4–5 GHz band is utilized by NATO members as a harmonized band for fixed and mobile military applications. Russia proposed to retain the coordination distances but without limits on IMT stations as protection was already provided by Article 9.21 of the RR (Youell, 2019b).

Accordingly, WRC-15 requested the ITU-R to conduct studies on the technical and regulatory conditions for the use of IMT in the 4800–4990 MHz frequency band in order to protect the aeronautical mobile service. However, no consensus had been reached by the ITU-R. During the second session of CPM-19, the issue was discussed, and it was recognized that the protection criteria was subject to review at WRC-19.

Prior to the conference, several RCC countries expressed the view that the 4800–4990 MHz band is the only alternative to C-band for early 5G deployment due to the need for spectrum refarming. Furthermore, the 4800–4900 MHz band would provide additional capacity and meet the market demand for 5G. Hence, the RCC countries proposed that WRC-19 should remove the hard power flux density (pdf) limit from Footnote RR 5.441B and allow countries to be added to the footnote in order to improve conditions for IMT system application in the 4800–4990 MHz band. The RCC countries also felt that the application of pdf level in addition to RR No 9.21 for the protection of the aeronautical mobile service (AMS) was redundant, as RR Article 8 already defined the need for protecting AMS stations and the regulatory practice of protection of AMS stations from IMT interference in other frequency bands. In addition, they requested that the agreement for application of IMT stations under No 9.21 should only be obtained from those countries using aeronautical stations for aeronautical data links (ADLs) in the frequency bands under No 5.442 (RRC, 2019b).

During the conference, the RCC countries proposed to exclude the pdf threshold criterion as it is not required to protect AMS and limit the use of IMT in the 4800–4990 MHz frequency bands (RRC, 2019a). However, CEPT, which includes NATO members, proposed to retain the regulatory and technical criteria in RR Article 5.441 (CEPT, 2019a). CITEL also supported retaining pdf protection limit along with any other appropriate technical criteria (CITEL, 2019).

Finally, due to strong lobbying by Russia with many other countries, NATO countries agreed to a compromise in the band in exchange for Russia's support for other agenda items in a 'package deal' (Marti, 2019c). More specifically, WRC-19 concluded that specific co-ordination distances between IMT and aircraft stations in the 4800–4825 MHz and 4835–4950 MHz frequency bands were required. In addition, the pdf limits in No. 5.441B, which is subject to review at WRC-23, would not apply to the following countries: Armenia, Brazil, Cambodia, China, Russian Federation, Kazakhstan, Lao P.D.R., Uzbekistan, South Africa, Vietnam and Zimbabwe (ITU-R, 2019f). This exclusion is extraordinary as it includes some countries which are unrelated, at least geographically, from specific conditions. For instance, countries such as Angola, Benin and Sudan are subject to a maximum pdf of IMT stations of -155 dB (W/(m^2· 1 MHz)), while other African countries such as South Africa and Zimbabwe are exempted from such condition (ITU-R, 2020b).

Therefore, it is not a surprise that the first item for WRC-23 to consider, based on the results of the ITU-R studies, are the possible measures to protect those AMS located in international airspace and waters from other stations located within national territories (ITU-R, 2019a).

10.3.3 Global Maritime Distress Safety System

One of the maim maritime issues that was discussed during the conference with a global influence was A.I. 1.8, which considered possible regulatory actions to support Global Maritime Distress Safety System (GMDSS) modernization and to support the introduction of additional satellite systems into GMDSS. WRC-19 Agenda Item 1.8 covered two separate items. The first was GMDSS modernization, while the second was the introduction of an additional satellite system into GMDSS. This is a result of the IMO decision to recognize a non-GSO MSS system, namely Iridium, operating in the 1616–1626.5 MHz frequency band that was expected to come into GMDSS operation in early 2020 (ITU-R, 2019b).

In the second item operator, iridium, which was recognized by IMO in 2018, was expected to cover areas not covered by the available GMDSS mobile-satellite services and to provide an alternative service provider for the maritime community (USA, 2018) The service would be capable of operating in the 1616–1626.5 MHz band (USA, 2018).

Agenda Item 1.8 highlighted the relationship between the ITU and other UN organizations. For GMDSS, IMO is responsible of defining operational needs, while the ITU determines the regulatory requirements. IMO invited WRC-19 to support the introduction of additional satellite systems into GMDSS before the start of 2020 while ensuring protection for those services using the recognized bands (Lim, 2019). During the conference, while the USA argued for larger bands, countries such as China and Russia declared their opposition and called for smaller bands (CEPT, 2019b). Ultimately, WRC-19 decided to support GMDSS modernization by including additional frequencies in the 415–526.5 kHz and 4–27.5 MHz bands for NAVDAT system under maritime mobile service allocations.

With regards the second issue, WRC-19 decided to allocate the 1621.35–1626.5 MHz band to the maritime mobile-satellite service ('MMSS') (space-to-earth) on a coprimary basis (DCCAE, 2019).

10.3.4 High-Altitude Platform Stations

High-altitude platform stations (HAPS) are one of the technologies that promises to deliver broadband services to a large area, especially in rural areas, providing additional capacity to existing systems or operating during disasters (ITU, 2019c). HAPS operates in the stratosphere, which is the layer of the earth's atmosphere that starts at 20 kms (ITU, 2019c).

Within the RR, a high-altitude platform station is defined as *'[a] station located on an object at an altitude of 20 to 50 km and at a specified, nominal, fixed point relative to the Earth'* (ITU-R, 2020a). In Article 5, there are several footnotes that determine the bands where HAPS can operate within a set of specific conditions (ITU-R, 2020b). Firstly, HAPS operates within the fixed service, and the wording used—'identification'—is similar to the one utilized for IMT. Additionally, identification does not provide priority to HAPS and indicates that the operation of HAPS must not cause harmful interference to existing services or claim protection from them. WRC-19 Agenda Item 1.14 considered the additional spectrum needs of gateway and fixed terminal links for HAPS to provide broadband connectivity in the fixed service (FS) (WRC-15).

Before the conference, ITU-R studies determined the total spectrum needs for HAPS systems to be in the range of 396 MHz (for lower capacity) to 2969 MHz (for higher capacity) for ground to HAPS platform links, and between 324 MHz (for lower capacity) to 1505 MHz (for higher capacity) for HAPS platform to ground links. This indicated that the existing HAPS identifications were insufficient (ITU-R, 2018b).

During the conference, some regional groups were sceptical about supporting the HAPS systems. For instance, ASMG did not support any additional identifications to applications of HAPS irrespective of results of ongoing studies under Agenda Items 1.6 and 1.13. It was agreed that HAPS would operate globally within the allocations to the fixed service in the 31–31.3 GHz, 38–39.5 GHz frequency bands. Furthermore, two bands (21.4–22 GHz and 24.25–27.5 GHz) were identified for HAPS use in the fixed service in Region 2. A compromise was necessary to avoid status changes in the 6440–6520 MHz and 27.9–28.2 GHz bands.

10.3.5 Radio Local Area Networks

A.I. 1.16 focused on issues related to wireless access systems including radio local area networks (WAS/RLAN) and allocated additional spectrum in accordance with

Resolution 239 (from WRC-15). This resolution acknowledged that that there had been considerable growth in the demand for WAS, with the bands studied allocated to different radiocommunication services such as aeronautical radio navigation service and the radiolocation service (ITU-R, 2015e). During the studies, the frequency range between 5150 and 5925 MHz was divided into five bands: A (5150–5250 MHz), B (5250–5350 MHz), C (5350–5470 MHz), D (5725–5850 MHz) and E (5850–5925 MHz). The regional groups differed in their positions, but there was an agreement on no change in the status for bands B, C and E (Marin, 2019).

WRC-19 eventually agreed to no change for three frequency bands (5250–5350 MHz, 5350–5470 MHz, 5850–5925 MHz) and modified Resolution 229 to allow outdoor usage for WAS/RLAN stations. In addition, it was decided to restrict the number of outdoor WAS/RLAN stations to two per cent of the total number of RLANS.

Two observations could be drawn from the discussions on Agenda Item 1.16. Firstly, it seems that there are different perspectives on the operations of unlicensed devices. On one hand, there is the USA which allows higher transmitted power and enforces less restrictive regulations on the operations of these devices. On the other hand, there were countries that were concerned about controlling these devices. This explains why, during the conference, there was a proposal to have a regional option that included 200 mW as a baseline worldwide option with further options in Region 2 and some countries in Regions 1 and 3 in the 5150–5250 MHz band (CEPT, 2019d). Secondly, the discussion also questions whether international regulations should play a role in regulating the use of services nationally, or if measures like the control of outdoor stations is applicable across different countries.

10.3.6 Transportation and Smart Cities

The railway radiocommunication systems that link the train with the trackside (RSTT) as well as Intelligent Transport Systems (ITS) were discussed during WRC-19. It was expected that these discussions would be finalized within the first week of the conference. However, the discussion of these two topics took much longer, lasting until the last couple of days of the conference due to the differences between participants regarding the way forward. According to A.I. 1.11, the conference needed to take action to facilitate harmonization, regionally or globally, so that trains and the trackside could communicate within the existing mobile service allocations. This reflected Resolution 263 from WRC-15 (ITU-R, 2015d).

ITU-R studies prior to the conference proposed three options (ITU-R, 2019b). The first was to make no changes to the RR except for the suppression of Resolution 236 (WRC-15), while the second was to add a new WRC resolution specifying frequency ranges for RSTT. The third option was to add a new WRC resolution without specifying frequency ranges for RSTT.

Before the conference, a number of ITU-R studies indicated that some countries in each of the three ITU regions had already designated the 5850–5925 MHz frequency band for the deployment of ITS (ITU-R, 2019b). With this in mind, three alternative approaches were suggested to satisfy the resolution from WRC-15. The first was to introduce no changes into the RR with ITS continuing to operate within existing mobile service allocations, with the harmonization of the relevant frequencies for ITS being achieved through ITU-R Recommendations and Reports. The second is similar to the first one but with the addition of a new WRC resolution to encourage countries to use 5850–5925 MHz for ITS. The third approach was to have a new WRC resolution to encourage countries to use the globally and regionally harmonized frequency bands for evolving ITS applications. The resolution would refer to Recommendation ITU-R M.2121, which recommended that several frequency bands within each region for use by current and future ITS applications (ITU-R, 2019d).

The discussions on ITS and RSTT continued throughout the conference due to the resistance of CEPT to including any new recommendations or resolutions in the RR addressing them. In particular, CEPT considered that the European measures for ITS and harmonization in the 5855–5925 MHz band were sufficient, and that any extra work could be developed through ITU-R recommendations. A similar European position was proposed regarding RSTT (CEPT, 2019a).

Eventually, WRC-19 issued a new resolution addressing transportation that only encourages countries when planning their RSTT to consider ITU–R study results to facilitate spectrum harmonization. It does, however, specify particular bands for RSTT harmonization, leaving instead the decision to national authorities to determine how much spectrum to make available for RSTT as well as the conditions for its use (ITU, 2019d; ITU-R, 2019g). In addition, WRC-19 issued a recommendation regarding ITS that suggested that each country should consider using the global or regional harmonized frequency bands for ITS when planning and deploying evolving these applications (ITU-R, 2019c). National authorities were also asked to take into account of coexistence issues between ITS and existing services (such as FSS earth stations).

10.3.7 Satellite Communications

WRC-19 included several agenda items related to satellites including spectrum for satellite-based broadband Internet access on moving platforms such as ships and trains, as well as regulating the deployment of large constellations of non-geostationary-satellite systems to prevent radio-frequency warehousing (Maniewicz, 2019a).

A.I. 1.5 considered the use of the 17.7–19.7 GHz (space-to-earth) and 27.5–29.5 GHz (earth-to-space) frequency bands by moving earth stations communicating with geostationary space stations in the fixed-satellite service (ITU-R, 2015c). The complexity of the issue comes from the fact that there are three types of ESIM—aeronautical, maritime and land—depending on the type of vehicle in which they are installed, and each type needs to protect exiting services while respecting their own operational conditions (ITU-R, 2019b).

During the conference, there were different opinions regarding the technical conditions that ESIM should comply with to protect terrestrial services (CEPT, 2019c). There was also some disagreement with respect to the pdf mask adopted for aeronautical ESIM and the examination characteristics of ESIMs with respect to the GSO FSS satellite network (CEPT, 2019d). WRC-19, in resolution CPM5/6, decided that '*transmitting aeronautical and maritime ESIM in the frequency band 27.5–29.5 GHz shall not cause unacceptable interference to terrestrial services to which the frequency band is allocated and operating in accordance with the Radio Regulations*' and '*transmitting land ESIM in the frequency band 27.5–29.5 GHz shall not cause unacceptable interference to terrestrial services in neighbouring countries to which the frequency band is allocated and operating in accordance with the Radio Regulations*'.

In addition, as the two bands (i.e. 17.7–19.7 GHz and 27.5–29.5 GHz) were already allocated to other radiocommunication services, several conditions were outlined in order not to cause unacceptable interference to services operating in neighbouring countries. The resolution also instructed the ITU Secretary-General to ensure that the Secretaries-General of the IMO and the International Civil Aviation Organization (ICAO) were informed (ITU, 2019c).

One critical agenda item regarding satellite services was A.I. 7, which considered possible changes in response to Resolution 86 of the Plenipotentiary Conference (Rev. Marrakesh, 2002), to frequency assignments for satellite networks as outlined by Resolution 86 (Rev.WRC-07). Resolution 86 requires that future WRC consider any proposals that seeks to improve the procedures associated with the RR for frequency assignments for space stations.

One of the important issues during WRC-19 was A.I. 7(A) that examined the use of frequency assignments to all non-GSO systems and considered a milestone-based approach for the deployment of non-GSO systems in specific frequency bands and services. The ITU-R studies on the issue focused on two topics (ITU-R, 2019b). The first was related to the concept of 'bringing into use'. There was an agreement that it could be achieved by the deployment of a single satellite into one of the notified orbital planes within seven years of the date of receipt of the advance publication of information (API) or request for coordination. However, there were four options with respect to the minimum period during which a satellite has to be maintained in a notified orbital plane.

The second topic focused on the details of the implementation of such a milestone-based approach for the deployment of non-GSO systems in specific frequency bands and services. The discussion of these issues during WRC-19 was arguably tense, primarily because the agreed date would impact on satellite operators. SpaceX, backed by its launching capabilities, and One Web, supported by a number of large investors, both sought shorter deadlines (Marti, 2019b).

A.I. 7 Issue A addressed the problem over NGSO fillings to the ITU that emerged as a result of the boom in the satellite industry. Companies such as SpaceX, OneWeb and Kuiper are planning to launch thousands of satellites, and hence need to submit a large number of satellite fillings to the ITU. The issue is that these fillings are usually based on speculation rather than accurate calculations, with the term 'paper satellite' frequently being used to describe them. While the ITU has a seven-year deadline for a satellite filing to be used, only one satellite is needed for a BIU (bring into use) to be realized—in other words, the thousands of filling by one company could be secured by the launch of just a single satellite (Youell, 2019a).

After a fraught discussion, it was decided that large constellations will need to deploy 10% of their constellation within two years after the end of the current regulatory period to bring them into use, half within five years and complete the deployment within seven years. Furthermore, it was decided that for those satellite systems whose regulatory period ended before 28 November 2022, they would be exempted from meeting the first milestone on the condition of reporting the current deployment and operational information by 1 April 2023 (ITU-R, 2020e).

One other important satellite agenda item was A.I.7 Issue E, which addressed RR Appendix 30B. This is a plan for fixed-satellite services in the 4500–4800 MHz, 6725–7025 MHz, 10.70–10.95 GHz, 11.20–11.45 GHz and 12.75–13.25 GHz frequency bands. More specifically, when a country wishes to convert its national allotment in RR Appendix 30B to assignments with characteristics different to the initial allocation, it will encounter several difficulties. It is, therefore, in practice, extremely difficult to successfully co-ordinate the change from national allotments to assignments whose characteristics differ from the initial allotment (ITU-R, 2019b). Studies prior to WRC-19 suggested establishing a series of special measures that could be applied once a submission had been received from a country with no frequency assignments in RR Appendix 30B. This would allow countries, especially developing ones, that do not have any assignments in the Appendix 30B to have privilege rules to access the list and develop a national system whose parameters were slightly different from their national allotment (ITU-R, 2019b). The conference decided that special procedures would be applied to process any submission from a country requesting that their allocation is modified, but the procedures can only be applied once (ITU-R, 2020f).

10.4 World Radiocommunication Conference 2023

Preparations are now well under way for the next WRC, which will convene in November 2023. Before addressing WRC-23, we need to explain the role of the CPM. The CPM holds two sessions, the first of which directly followed WRC-19 to organize the preparatory studies for WRC-23. The second one will be held around six months before WRC-23 to finalize and approve its CPM report to the conference.

The importance of such a report is because it often highlights the main differences in views and sometimes enables the convergence of these different views around proposed solutions (Maniewicz, 2019b). In other words, it offers the possibility for compromises to be reached. They are also important because not all ITU members attend the study group meetings, with the consequence that the CPM report is a key document that enables them to learn, before the WRC, about the various studies that have been conducted (Awadi, 2019).

The agenda of WRC-23 accommodates almost 30 agenda items (ITU-R, 2020h). CPM-23 grouped them into five categories: fixed, mobile and broadcasting issues; aeronautical and maritime issues; science issues; satellite issues; and general issues – see Appendix 1. The main agenda items for WRC-23, however, can be found in ITU-R (2019a). It is important to recognise that these agenda items were selected and agreed upon from tens of proposals. At WRC-19, 72 proposals covering 25 different topics were submitted that would become new agenda items for WRC-23 and WRC-27. Various factors determine which proposals will be accepted, and at which conference—WRC-23 or WRC-27—they will be addressed.

During WRC-19, it was thought that some particular Working Parties (e.g. 4A and 5B) would be overloaded with work if all the potential proposals were accepted. Several topics received multiple proposals—for instance, there were 14 proposals relating to IMT, five for GMDSS and five for space–space FSS. The large number of proposals for WRC-23, coupled with the lack of urgency for some of them, resulted in a decision to defer consideration of some proposals to WRC-27. Finally, it is worth noting that the proposals reveal the differing interests of regional organizations. For example, while CEPT submitted a large number of agenda items for future conferences, none of their proposals were for additional IMT identification(s).

A.I. 1.3 requires careful examination as it only considers the primary allocation in part of the 3600–3800 MHz band allocated to mobile services without IMT identification. To understand the issue, it important to highlight that prior to WRC-19, several European countries assigned, or planned to assign, the C-band to 5G services (El-Moghazi & Whalley, 2019). Furthermore, several Arab countries announced that they would be using the 3.4–3.8 GHz band for IMT before WRC-19 (El-Moghazi & Whalley, 2019). During the conference, ASMG submitted a proposal to study elevating the current secondary allocation of the mobile service in the 3600–3800 MHz frequency band to primary allocation within Region 1, and identifying the band for the future development of IMT. This was also backed by many African countries. However, the IMT identification in the 3.6–3.8 GHz band had previously been discussed in WRC-07 and WRC-12 without agreement. Having countries in Region 1 agree to operate IMT systems in bands allocated to mobile service only without IMT identification raised a lot of questions regarding the importance of future IMT identification.

Regarding the UHF band, A.I. 1.5 outlined the need *'to review the spectrum use and spectrum needs of existing services in the frequency band 470–960 MHz in Region 1 and consider possible regulatory actions in the frequency band 470–694 MHz in Region 1'* (ITU-R, 2020h). The ASMG and CEPT groups supported the study of the UHF band for WRC-23. On the other hand, a few Arab countries (e.g. Egypt and UAE) sought the modification of the AI to explicitly mention the possibility of IMT identification in the 470–694 MHz band. CEPT supported retaining the AI as it was agreed at WRC-15 (RSPG, 2018). In contrast, ATU countries highlighted the fact that DTT broadcasting is important for the majority of African countries (SFCG, 2019b). RCC countries explicitly objected to the inclusion of the AI in WRC-23, as in Region 1 the 470–694 MHz frequency band is intensively used by existing services including broadcasting (RRC, 2019a). It was eventually agreed that A.I. 1.5

is based on a finely balanced compromise that was reached and agreed at WRC-15 and changing any aspect of the agreement would raise a lot of objections. Therefore, the A.I. was not modified to reflect that all radiocommunication services will be reviewed. This did not imply IMT identification.

Although A.I. 1.2 included several bands in the studies of IMT identification, and is considered as a continuation of discussions undertaken at WRC-15 and WRC-19, the issue of IMT identification in the 6 GHz band gained in a lot of attention. In particular, before WRC-19, several US-based companies (e.g. Facebook, Apple, Cisco, Microsoft) demanded that the FCC resist the plans of mobile equipment manufacturers (e.g. Ericsson and Huawei) to study the 6 GHz for IMT. This is because the band is critical to the future of Wi-Fi. They also highlighted that the proposal of IMT in the 6 GHz band is supported by a handful of countries such as China and opposed by others across Asia and Europe (FCC, 2020). Meanwhile, the FCC considered expanding unlicensed use of the 5.925–7.125 GHz band (6 GHz band) while protecting the incumbent licensed services (e.g. point-to-point microwave links and FSS) that operate in this spectrum (FCC, 2020). In the EU, the 5925–6425 MHz band was considered for WAS/RLANs while protecting other radio services and applications currently in use (ECC, 2019).

During WRC-19, different proposals emerged regarding the future study of IMT in the 6 GHz band. The RCC countries supported 6525–7100 MHz, while the Arab countries did not support specific frequencies but instead called for studying the 6–24 GHz band range. Unlike the Asian countries, who agreed to studying only 7025–7125 MHz, China, encouraged by Huawei, supported the whole of the 5925–7125 MHz band (Chairman Working Group 6B, 2019). The APT region, which includes China, did not obtain a consensus before the conference due to the differences in the requirements between countries and their competing industrial interests (SFCG, 2019b). Therefore, it was only China, backed by Huawei, that was interested in studying the whole 6 GHz band for IMT. Eventually, it was decided to split the 6 GHz band between IMT and unlicensed application so that the lower band (5925—6425 MHz) was retained for unlicensed applications and the upper band (6425—7125 MHz) to be studied for IMT at WRC-23 (Marti, 2019d).

The final relevant issue for our purposes was A.I. 9.1.C, which studied the use of IMT system for fixed wireless broadband as outlined in Resolution 175 (ITU-R, 2020g). The frequency bands were allocated to the fixed services on a primary basis, which was highly surprising as it would mean utilizing IMT systems for fixed applications rather than the traditional mobile. During WRC-19, there was some resistance to the proposal as it focused on specific frequency bands. The proposal initially focused on the 10.7–11.7 GHz frequency bands which were allocated to fixed, mobile and fixed-satellite services on a primary basis across all the three ITU Regions (Bahrain (Kingdom of) et al., 2019). The merits of the proposal, according to those supporting it, were that the 10.7–11.7 GHz band is already utilized by point-to-multipoints and point-to-point fixed proprietary technologies with the consequence that IMT would provide a reliable alternative to them. During the last days of WRC-19, the plenary decided to remove the 10.7–11.7 GHz band from the agenda item and to study the issue of IMT operation in the fixed service (Marti, 2019d).

10.5 World Radiocommunication Conference 2027

One may wonder why Member States or sector members are so keen to include topics for discussion in WRC-27, a conference to be held in eight years' time. Including a topic in a future WRC agenda indicates interest in a certain frequency band or wireless system. In addition, some studies may take a considerable time to complete, especially if they are based on monitoring data. Furthermore, those frequency bands mentioned in agenda items provide the industry with a degree of certainty so that companies can develop their business models, services or products. However, the presence of an agenda item could be considered as a threat. This explains the widespread refusal by countries to have any agenda item revising the situation in the UHF band. Similarly, the 10.7–11.7 GHz bands were omitted from the proposal to study IMT for fixed application in AI 9.1.C of WRC-23. The reason is the uncertainty for the broadcasting industry in the first case due to where DTT operates, and the satellite industry in the second case where BSS operates.

The agenda of WRC-27 contains a number of interesting topics, but none of them address the issue of additional frequencies for IMT—see Appendix 2. This could be due to the influence of the substantial additional frequency identification for IMT in WRC-19, or because of the uncertainty related to future requirements of advanced systems of 5G, and, perhaps, 6G. A closer examination of the different items provides us with a hint into the future of wireless communications as well as the focus of the wireless industry in the coming decade. The topics are diverse, including ESIMS, FSS, space weather sensors, space-to-space links among non-geostationary and geostationary satellites operating in the mobile-satellite service, VHF maritime frequencies, EESS (earth-to-space) allocation and narrowband mobile-satellite systems.

10.6 Summary

In this chapter, we have highlighted the main outputs of the recent RA-19 and WRC-19, which were both held by the end of 2019, and explored the different agenda items for WRC-23 and WRC-27. Radio Assembly 2019 was responsible for the structure, programme and approval of studies, some of which informed the following WRC. WRC-19 was the first one in 20 years to be held outside of Switzerland, addressing several significant technical issues such as candidate 5G frequencies and issues of political nature such as the assistance and support to Palestine in developing its 3G and 4G systems. This highlighted that while the ITU may be a technical organization, it is not immune from politics, as did the discussion of IMT identification in the 4800–4900 MHz bands.

This chapter has also illustrated the needs of companies to obtain recognition and frequencies from ITU-R. Iridium sought both to become an alternative service provider focusing on maritime communities. Perhaps a surprising theme to emerge from the chapter, especially given the emphasis on consensus at the ITU, has been the divergent views on spectrum management issues. There were different views regarding license-exempt systems operating outdoors as well as the identification of IMT within the 4800–4900 MHz bands. Some of these differences were resolved, with compromises being agreed, while others were effectively postponed to future WRC.

The agenda for WRC-23, on which work has already begun, is broad. It ranges from the primary allocation of part of the 3600–3800 MHz band to mobile services without IMT identification to a review of spectrum needs and uses in the UHF band. When combined with the preliminary planning for WRC-27 that has commenced, not only does the need to continually manage and refine the allocation of spectrum to reflect technical change on the one hand and changing use patterns on the other come need to occur but questions around the suitability of existing procedures need to be raised.

Appendix 1: Agenda for the 2023 World Radiocommunication Conference

1.1	To consider, based on the results of the ITU-R studies, possible measures to address, in the frequency band 4800–4990 MHz, protection of stations of the aeronautical and maritime mobile services located in international airspace and waters from other stations located within national territories, and to review the pdf criteria in No. 5.441B in accordance with Resolution 223 (Rev.WRC-19)
1.2	To consider identification of the frequency bands 3300–3400 MHz, 3600–3800 MHz, 6425–7025 MHz, 7025–7125 MHz and 10.0–10.5 GHz for International Mobile Telecommunications (IMT), including possible additional allocations to the mobile service on a primary basis, in accordance with Resolution 245 (WRC-19)
1.3	To consider primary allocation of the band 3600–3800 MHz to mobile service within Region 1 and take appropriate regulatory actions, in accordance with Resolution 246 (WRC-19)
1.4	To consider, in accordance with Resolution 247 (WRC-19), the use of high-altitude platform stations as IMT base stations (HIBS) in the mobile service in certain frequency bands below 2.7 GHz already identified for IMT, on a global or regional level
1.5	To review the spectrum use and spectrum needs of existing services in the frequency band 470–960 MHz in Region 1 and consider possible regulatory actions in the frequency band 470–694 MHz in Region 1 on the basis of the review in accordance with Resolution 235 (WRC-15)
1.6	To consider, in accordance with Resolution 772 (WRC-19), regulatory provisions to facilitate radiocommunications for sub-orbital vehicles
1.7	To consider a new aeronautical mobile-satellite (R) service (AMS(R)S) allocation in accordance with Resolution 428 (WRC-19) for both the Earth-to-space and space-to-Earth directions of aeronautical VHF communications in all or part of the frequency band 117.975–137 MHz, while preventing any undue constraints on existing VHF systems operating in the AM(R)S, the ARNS, and in adjacent frequency bands

(continued)

(continued)

1.8	To consider, on the basis of ITU R studies in accordance with Resolution 171 (WRC-19), appropriate regulatory actions, with a view to reviewing and, if necessary, revising Resolution 155 (Rev.WRC-19) and No. 5.484B to accommodate the use of fixed-satellite service (FSS) networks by control and non-payload communications of unmanned aircraft systems
1.9	To review Appendix 27 of the Radio Regulations and consider appropriate regulatory actions and updates based on ITU R studies, in order to accommodate digital technologies for commercial aviation safety-of-life applications in existing HF bands allocated to the aeronautical mobile (route) service and ensure coexistence of current HF systems alongside modernized HF systems, in accordance with Resolution 429 (WRC-19)
1.10	To conduct studies on spectrum needs, coexistence with radiocommunication services and regulatory measures for possible new allocations for the aeronautical mobile service for the use of non-safety aeronautical mobile applications, in accordance with Resolution 430 (WRC-19)
1.11	To consider possible regulatory actions to support the modernization of the Global Maritime Distress and Safety System and the implementation of e navigation, in accordance with Resolution 361 (Rev.WRC-19)
1.12	To conduct, and complete in time for WRC 23, studies for a possible new secondary allocation to the Earth exploration-satellite (active) service for spaceborne radar sounders within the range of frequencies around 45 MHz, taking into account the protection of incumbent services, including in adjacent bands, in accordance with Resolution 656 (Rev.WRC-19)
1.13	To consider a possible upgrade of the allocation of the frequency band 14.8–15.35 GHz to the space research service, in accordance with Resolution 661 (WRC-19)
1.14	To review and consider possible adjustments of the existing or possible new primary frequency allocations to EESS (passive) in the frequency range 231.5–252 GHz, to ensure alignment with more up-to-date remote-sensing observation requirements, in accordance with Resolution 662 (WRC-19)
1.15	To harmonize the use of the frequency band 12.75–13.25 GHz (Earth-to-space) by earth stations on aircraft and vessels communicating with geostationary space stations in the fixed-satellite service globally, in accordance with Resolution 172 (WRC-19)
1.16	To study and develop technical, operational and regulatory measures, as appropriate, to facilitate the use of the frequency bands 17.7–18.6 GHz and 18.8–19.3 GHz and 19.7–20.2 GHz (space-to-Earth) and 27.5–29.1 GHz and 29.5–30 GHz (Earth-to-space) by non-GSO FSS earth stations in motion, while ensuring due protection of existing services in those frequency bands, in accordance with Resolution 173 (WRC-19)
1.17	To determine and carry out, on the basis of the ITU-R studies in accordance with Resolution 773 (WRC-19), the appropriate regulatory actions for the provision of inter-satellite links in specific frequency bands, or portions thereof, by adding an inter-satellite service allocation where appropriate
1.18	To consider studies relating to spectrum needs and potential new allocations to the mobile-satellite service for future development of narrowband mobile-satellite systems, in accordance with Resolution 248 (WRC-19)
1.19	To consider a new primary allocation to the fixed-satellite service in the space-to-Earth direction in the frequency band 17.3–17.7 GHz in Region 2, while protecting existing primary services in the band, in accordance with Resolution 174 (WRC-19)

(continued)

(continued)

9.1.a	In accordance with Resolution 657 (Rev.WRC-19), review the results of studies relating to the technical and operational characteristics, spectrum requirements and appropriate radio service designations for space weather sensors with a view to describing appropriate recognition and protection in the Radio Regulations without placing additional constraints on incumbent services
9.1.b	Review of the amateur service and the amateur-satellite service allocations in the frequency band 1240–1300 MHz to determine if additional measures are required to ensure protection of the radionavigation-satellite (space-to-Earth) service operating in the same band in accordance with Resolution 774 (WRC-19)
9.1.c	Study the use of International Mobile Telecommunication system for fixed wireless broadband in the frequency bands allocated to the fixed services on primary basis, in accordance with Resolution 175 (WRC-19)
9.1.d	Protection of EESS (passive) in the frequency band 36–37 GHz from non-GSO FSS space stations

Source ITU-R (2020h)

Appendix 2: Preliminary Agenda Items for WRC-27[a]

2.1	To consider, in accordance with Resolution 663 (WRC-19), additional spectrum allocations to the radiolocation service on a co-primary basis in the frequency band 231.5–275 GHz and identification for radiolocation applications in frequency bands in the range 275–700 GHz for millimetre and sub-millimetre wave imaging systems
2.2	Study and develop technical, operational and regulatory measures, as appropriate, to facilitate the use of the frequency bands 37.5–39.5 GHz (space-to-Earth), 40.5–42.5 GHz (space-to-Earth), 47.2–50.2 GHz (Earth-to-space) and 50.4–51.4 GHz (Earth-to-space) by aeronautical and maritime earth stations in motion communicating with geostationary space stations in the fixed-satellite service, in accordance with Resolution 176 (WRC-19)
2.3	To consider the allocation of all or part of the frequency band [43.5–45.5 GHz] to the fixed-satellite service, in accordance with Resolution 177 (WRC-19)
2.4	The introduction of pdf and e.i.r.p. limits in Article 21 for the frequency bands 71–76 GHz and 81–86 GHz in accordance with Resolution 775 (WRC-19)
2.5	The conditions for the use of the 71–76 GHz and 81–86 GHz frequency bands by stations in the satellite services to ensure compatibility with passive services in accordance with Resolution 776 (WRC-19)
2.6	To consider regulatory provisions for appropriate recognition of space weather sensors and their protection in the Radio Regulations, taking into account the results of ITU-R studies reported to WRC-23 under agenda item 9.1 and its corresponding Resolution 657 (Rev.WRC-19)
2.7	To consider the development of regulatory provisions for non-geostationary fixed-satellite system feeder links in the frequency bands 71–76 GHz (space-to-Earth and proposed new Earth-to-space) and 81–86 GHz (Earth-to-space), in accordance with Resolution 178 (WRC-19)

(continued)

(continued)

2.8	To study the technical and operational matters, and regulatory provisions, for space-to-space links in the frequency bands [1525–1544 MHz], [1545–1559 MHz], [1610–1645.5 MHz], [1646.5–1660.5 MHz] and [2483.5–2500 MHz] among non-geostationary and geostationary satellites operating in the mobile-satellite service, in accordance with Resolution 249 (WRC-19)
2.9	To consider possible additional spectrum allocations to the mobile service in the frequency band 1300–1350 MHz to facilitate the future development of mobile-service applications, in accordance with Resolution 250 (WRC-19)
2.10	To consider improving the utilization of the VHF maritime frequencies in Appendix 18, in accordance with Resolution 363 (WRC-19)
2.11	To consider a new EESS (Earth-to-space) allocation in the frequency band 22.55–23.15 GHz, in accordance with Resolution 664 (WRC-19)
2.12	To consider the use of existing IMT identifications in the frequency range 694–960 MHz by consideration of the possible removal of the limitation regarding aeronautical mobile in the IMT for the use of IMT user equipment by non-safety applications, where appropriate, in accordance with Resolution 251 (WRC-19)
2.13	To consider a possible worldwide allocation to the mobile-satellite service for the future development of narrowband mobile-satellite systems in frequency bands between the range [1.5–5 GHz], in accordance with Resolution 248 (WRC-19)

Notes

[a]Some agenda items have in their wording square brackets around certain frequency bands which indicates that WRC-23 will consider and review the inclusion of these frequency bands with square brackets and decide, as appropriate

Source ITU-R (2020i)

References

ASMG. (2019a). *Arab states common proposals. Common proposals for the work of the conference* WRC-19. ITU.

ASMG. (2019b). *Arab states common proposals. Common proposals for the work of the conference. Addendum 25* WRC-19.

Awadi, K. E. (2019). From conference preparatory meeting to WRC-19. *ITU News*(5). Retrieved 30 August 2020, from www.itu.int

Bahrain (Kingdom of), Kuwait (State of), Tunisia, & United Arab Emirates. (2019). *Proposals for the work of the conference—Agenda item 10* WRC-19. ITU.

CEPT. (2019a). *Proposals for the work of the conference* WRC-19. ITU.

CEPT. (2019b). *Report of the final week of the WRC-19.* Retrieved from www.cept.org

CEPT. (2019c). *Report of the first week of the WRC-19.* Retrieved from www.cept.org

CEPT. (2019d). *Report of the second week of the WRC-19.* Retrieved from www.cept.org

Chairman Working Group 6B. (2019). *Consideration of proposals relating to agenda item 10* WRC-19. ITU.

Chairman Radiocommunication Study Group 7. (2019). *Chairman report* radio assembly—2019. ITU.

Chairman Study Group 5. (2019). *Summary record of the fifteenth meeting of study group 5* study group 5.

CITEL. (2019). *Proposals for the work of the conference* WRC-19, ITU.

DCCAE. (2019). Outcome of WRC-19. Retrieved 30 September 2020, from https://www.dccae.gov.ie
ECC. (2019). Europe prepares to harmonise the 6 GHz spectrum band for radio local area networks. *ECC Newsletter* (August). Retrieved 30 September 2020, from http://apps.cept.org
El-Moghazi, M., & Wahelly, J. (2019). *IMT spectrum identification: Obstacle for 5G deployments* TPRC47: 47th Research Conference on Communications, Information and Internet Policy.
FCC. (2020). *Unlicensed use of the 6 GHz band*. Retrieved from https://www.govinfo.gov
Global Satellite Coalition (GSC). (2019). Maintaining and expanding spectrum for satellite communications. *ITU News*. Retrieved 12 November 20, from www.itu.int
IDate & Plum. (2019). Study on using millimetre waves bands for the deployment of the 5G ecosystem in the union. Retrieved 1 August 2021, from https://op.europa.eu/
ITU. (2014). *Resolution 185: Global flight tracking for civil aviation* PP-14. ITU.
ITU. (2015a). Press release: Israeli-Palestinian agreement reached on granting Palestinian operators "exclusive frequencies". Retrieved 30 September 2020, from www.itu.int
ITU. (2015b). *Radio spectrum allocated for global flight tracking*. Retrieved 20 September 2020, from www.itu.int
ITU. (2019a). *Convention of the international telecommunication union* (Collection of the basic texts of the ITU adopted by the plenipotentiary conference (Ed 2019)). ITU.
ITU. (2019b). *ITU radiocommunication assembly sets stage for further evolution of innovative digital technologies*. Retrieved 30 September 20, from www.itu.int
ITU. (2019c). Key outcomes of the world radiocommunication conference 2019. *ITU News*. Retrieved 30 September 2020, from www.itu.int
ITU. (2019d). Key outcomes of WRC-19 in brief. *ITU News*(6). Retrieved 30 September 2020, from www.itu.int
ITU. (2019e). WRC-19 identifies additional frequency bands for 5G. *ITU News*. Retrieved 30 September 20, from www.itu.int
ITU-R. (2012). Provisional final acts—World radiocommunication conference (WRC-12). ITU.
ITU-R. (2015a). Handbook on national spectrum management. ITU.
ITU-R. (2015b). *Resolution 12 (Rev.WRC-15): Assistance and support to palestine* WRC-15, ITU.
ITU-R. (2015c). *Resolution 158: Use of the frequency bands 17.7–19.7 GHz (space-to-Earth) and 27.5–29.5 GHz (earth-to-space) by earth stations in motion communicating with geostationary space stations in the fixed-satellite service* WRC-15, ITU.
ITU-R. (2015d). Resolution 236: Railway radiocommunication systems between train and trackside WRC-15, ITU.
ITU-R. (2015e). Resolution 239 (WRC-15) studies concerning wireless access systems including radio local area networks in the frequency bands between 5150 MHz and 5925 MHz, WRC-15, ITU.
ITU-R. (2015f). *M.1036-5 (10/2015): Frequency arrangements for implementation of the terrestrial component of international mobile telecommunications (IMT) in the bands identified for IMT in the radio regulations*. ITU.
ITU-R. (2016a). Article 5: Frequency allocation. In *RR-2016*. ITU.
ITU-R. (2016b). Resolution 238: Studies on frequency-related matters for international mobile telecommunications identification including possible additional allocations to the mobile services on a primary basis in portion(s) of the frequency range between 24.25 and 86 GHz for the future development of international mobile telecommunications for 2020 and beyond. In *Radio regulations*. ITU.
ITU-R. (2016c). Resolution 809: Agenda for the 2019 world radiocommunication conference. In *Radio regulations*. ITU.
ITU-R. (2018a). ITU-R recommendation M.2003-2: Multiple gigabit wireless systems in frequencies around 60 GHz. In *ITU-R recommendations M-Series*. ITU.
ITU-R. (2018b). Report ITU-R F.2438-0: Spectrum needs of high-altitude platform stations broadband links operating in the fixed service. In *ITU-R reports F-Series*. ITU.

ITU-R. (2019a). *Administrative circular CA/251: Results of the first session of the conference preparatory meeting for WRC-23 (CPM23-1)*. Retrieved from www.itu.int
ITU-R. (2019b). *CPM report on technical, operational and regulatory/procedural matters to be considered by the 2019 world radiocommunication conference* CPM-19, ITU.
ITU-R. (2019c). *Recommendation 208: (WRC-19) harmonization of frequency bands for evolving Intelligent transport systems applications under mobile-service allocations* WRC-19, ITU.
ITU-R. (2019d). Recommendation M.2121-0: Harmonization of frequency bands for Intelligent transport systems in the mobile service. In *ITU-R recommendations M-Series*. ITU.
ITU-R. (2019e). Resolution 71: Role of the radiocommunication sector in the ongoing development of television, sound and multimedia broadcasting. In *ITU-R resolutions*. ITU.
ITU-R. (2019f). *Resolution 223 (REV. WRC-19): Additional frequency bands identified for international mobile telecommunications* WRC-19, ITU.
ITU-R. (2019g). *Resolution 240 (WRC-19): Spectrum harmonization for railway radiocommunication systems between train and trackside within the existing mobile-service allocation* WRC-19, ITU.
ITU-R. (2019h). *Resolution 241: (WRC-19) use of the frequency band 66–71 GHz for international mobile telecommunications and coexistence with other applications of the mobile service* WRC-19, ITU.
ITU-R. (2019i). *Summary report of the sixth plenary meeting of the radiocommunication assembly* radio assembly—2019, ITU.
ITU-R. (2020a). Article 1: Terms and definitions. In *Radio regulations*. ITU.
ITU-R. (2020b). Article 5: Frequency allocations. In *Radio regulations*. ITU.
ITU-R. (2020c). Resolution 12: (Rev.WRC-19) assistance and support to Palestine. ITU.
ITU-R. (2020d). Resolution 26: (Rev.WRC-19) footnotes to the table of frequency allocations in article 5 of the radio regulations. In *Radio Regulations*. ITU.
ITU-R. (2020e). Resolution 35 (WRC-19): A milestone-based approach for the implementation of frequency assignments to space stations in a non-geostationary-orbit satellite system in specific frequency bands and services In *RR 2020*. ITU.
ITU-R. (2020f). Resolution 170 (WRC-19): Additional measures for satellite networks in the fixed-satellite service in frequency bands subject to Appendix 30B for the enhancement of equitable access to these frequency bands In *RR 2020*. ITU.
ITU-R. (2020g). Resolution 175 (WRC-19): Use of international mobile telecommunications systems for fixed wireless broadband in the frequency bands allocated to the fixed service on a primary basis In *RR-2020*. ITU.
ITU-R. (2020h). Resolution 811: (WRC-19) agenda for the 2023 world radiocommunication conference In *RR-2020*. ITU.
ITU-R. (2020i). Resolution 812: (WRC-19) preliminary agenda for the 2027 world radiocommunication conference. In *RR-2020*. ITU.
ITU-R BR Director. (2019). *Report of the director on the activities of the radiocommunication sector* WRC-19, ITU.
Jeanty, L. (2019). The radio regulations board and WRC-19. *ITU News*(5). Retrieved from www.itu.int
Kraemer, M. (2019). *Sessions 1 & 2 IMT related issues, agenda item (AI) 1.13*, 3rd interregional workshop on WRC-19 preperations. Retrieved from www.itu.int
Lim, K. (2019). Maritime communications — safeguarding the spectrum for maritime services. *ITU News*(5).
LS Telecom. (2019). *Analysis of the world-wide licensing and usage of IMT spectrum*. Retrieved from https://www.esoa.net
Maniewicz, M. (2019a). Satellite communications—An essential link for a connected world. *ITU News*(2). Retrieved 13 October 2020, from www.itu.int
Maniewicz, M. (2019b). WRC-19: Enabling global radiocommunications for a better tomorrow. *ITU News*(5). Retrieved from www.itu.int

Marin, H. (2019). *WRC-19 Agenda item 1.16* 3rd ITU Inter-regional workshop on WRC-19 preparation. Retrieved from www.itu.int

Marti, M. R. (2019a). RA-19 paves the way for a global broadcasting standard. *PolicyTracker*. Retrieved 30 September 2020, from www.policytracker.com

Marti, M. R. (2019b). WRC-19 discusses NGSO legalities amid lobbying battle. *PolicyTracker*. Retrieved 30 August 2020, from www.policytracker.com

Marti, M. R. (2019c). WRC-19 identifies 4.8 GHz for IMT in surprise move. *PolicyTracker*. Retrieved 2 February 2020, from www.policytracker.com

Marti, M. R. (2019d). WRC-23: What you need to know today. *PolicyTracker*. Retrieved 30 September 2020, from www.policytracker.com

RRC. (2019a). *Proposals for the work of the conference* WRC-19.

RRC. (2019b). *Status of RCC's proposals to WRC-19 and RA-19* 3rd ITU inter-regional workshop on WRC-19 preparation. Retrieved from www.itu.int

RSPG. (2018). *Final RSPG opinion on the ITU-R world radiocommunication conference 2019*. Retrieved from https://rspg-spectrum.eu

Russia. (2019). *Proposals in relation to agenda Item 1.13 concerning conditions for IMT use in the 24.25–27.5 GHz band* WRC-19, ITU.

SFCG. (2019a). *SFCG objectives for WRC-19* 3rd interregional workshop on WRC-19 preperations. Retrieved from www.itu.int

SFCG. (2019b). *Status of regional proposals and positions for WRC-19*. Retrieved from https://www.sfcgonline.org/

USA. (2018). *Revision to inter-American proposal on WRC-19 agenda item 1.8* CITEL preperation meeting for WRC-19.

Viola, C. (2020). The implications of WRC-19 for 5G. *PolicyTracker*. Retrieved 30 July 2021, from www.policytracker.com

WMO. (2019). *WMO position on WRC-19 agenda* 3rd interregional workshop on WRC-19 preperations.

Youell, T. (2019a). Mega-NGSO filing arrives at ITU as concerns are raised over current system's sustainability. *PolicyTracker*. Retrieved 30 August 2020, from www.policytracker.com

Youell, T. (2019b). Russia proposes 5G transmission in NATO band. *PolicyTracker*. Retrieved 30 August 2020, from www.policytracker.com

Chapter 11
The Way Forward

> *Global technological leaders, such as the U.S., will continue to innovate outside and without input from the ITU and its many nation states. This will, in turn, make the ITU and the WRC process less relevant.... Structural reforms need to be enacted to ensure that the ITU remains technology neutral and focused on its core mission... Failure to proceed along this path is likely to lead to calls for the U.S. to defund the ITU in whole or in part, which would likely fracture the organization and lead to its functional demise.*
>
> Michael O'Rielly, FCC Commissioner (2016)

11.1 Introduction

This book's main argument is that it is time to revise the more than 100-year-old radio regulations that have, in several instances, blocked technology innovation and restricted the sovereignty of countries over radio spectrum, one of the most important resources today. Recent WRCs have highlighted some of the pitfalls in ITU-R decision-making procedures, with countries trading positions based on politics and other unrelated issues, which could block the interests of other countries even if they are geographically far away. This could have a significant impact on the relevance of WRCs to the wider wireless community, both today and in the future. In general, the main problems with the RR are twofold: the restrictive elements within them and the paradigm that has been created over the course of the last century.

11.2 Interference Management

Interference seems to be at the centre of any spectrum management discussion regardless of the geographical scale, and, as we have elaborated in previous chapters, the main rationale behind spectrum service allocation is the management of interference between radio stations (Louis & Mallalieu, 2007). How interference is embedded

M. A. El-Moghazi and J. Whalley, *The International Radio Regulations*,
https://doi.org/10.1007/978-3-030-88571-7_11

within the current RR? Harmful interference is defined, according to the ITU-R RR, as *'interference which endangers the functioning of a radio navigation service or of other safety services or seriously degrades, obstructs, or repeatedly interrupts a radiocommunication service operating in accordance with Radio Regulations'* (ITU-R, 2008b). The interference-resolving procedures in the RR are initiated when the victim country completes a harmful interference report (ITU-R, 2012a). The rest of the procedures are based on the goodwill of the involved countries. Most importantly, the ITU-R Bureau (BR) can assist when asked to determine the source of interference. However at the end of the day, the role of the BR is limited to forwarding its conclusions and recommendations to the country reporting the case of harmful interference, and to the country believed to be responsible for the source of harmful interference together with a request for prompt action (ITU-R, 2008b).

Accordingly, resolving any dispute will refer to the values of permissible interference and accepted interference based on the ITU-R Recommendations (Indepen, 2001). However, this degradation of service or limitations of interference are not accurately defined or quantified in many cases in the RR, and as the ITU-R recommendations do not cover all cases of interference and thus some solutions cannot be enforced. Additionally, the ITU does not have the authority to enforce its rules or bring sanctions to those who seek to operate outside its agreed procedures (Cronk, 2011). More specifically, while these recommendations are usually based on worst-case scenarios in terms of interference probability, they are not enforced as they are 'merely' recommendations (Indepen, 2001).

Even these recommendations do not contain any provisions related to the spill over between neighbouring countries, and there is no recommendation dealing with the issue, which is mostly left to multi/biliteral agreements. This is, in fact, the main problem with international interference management, which focuses on the ex ante procedures of interference management because the ex post procedures are neither strict nor enforced effectively (RSPG, 2013). Moreover, the ITU-R does not have clear mechanisms for identifying sources of interferences or determining whether it is unintended or due to jamming. Finally, there are no measures for the ITU-R to investigate the actual spectrum use (Struzak, 2007b).

Examining RR Article 4.4, which allows operating on a non-interference basis, reveals that the application of such an article provides no cross-border protection against interference (Louis, 2011). Thus, even if operating according to the RR Article 4.4 is reasonable in terms of transmitted power and used antennas, it is not protected against interference from neighbouring countries. This, of course, does not provide any type of encouragement to the adoption of flexible spectrum use models such as opportunistic access. Hence, while Article 4.4 allows for non-interference operations, for many countries such an article is not used in practice or is not reliable for commercial or public sector usage as usually stated through ITU-R discussions. More specifically, when a country requires a new allocation and a neighbouring country calls for operating on a non-interference basis according to Article 4.4, this implies allowing nothing in practice.

Similarly, Article 9.21 requires that before a country notifies the ITU-R Bureau, or brings into use a frequency assignment for any station of a service, it must seek the agreement of other countries is included in a footnote to the RR table of frequency allocations, and it shall seek the agreement of the affected countries (ITU-R, 2012b). Article 9.21 is perceived to be similar to Article 4.4 in terms of discouraging more sharing of spectrum in several ways. Not only does it indicate no interference of existing radiocommunication services or neighbouring countries but it is also becoming increasingly unpopular as it does not guarantee protection against interference from the prospective of some countries. In fact, the practice of Article 9.21 application indicates that it is not much used in practice (World Radiocommunication Report, 2015b).

There also appears to be a link between harmonization and interference where one of the drivers of the former is to avoid the latter. More specifically, if a country does not have a harmonized allocation with its neighbouring countries, it does not have protection against interference and thus has a greater probability of being affected by interference from these neighbouring countries. One example is the global support for mobile harmonization in the 700 MHz band across the three ITU-R regions. While this is mainly driven by the benefits of economies of scales and roaming, it is also related to the fear of interference from broadcasters.

The link between harmonization and interference is also apparent in the discussions related to the future use of the UHF band in Europe. The high-level group, which was established in 2014 by the European Commission to examine the future of 470–790 MHz band (Youell, 2014), concluded that linear TV viewing will remain dominant for the foreseeable future. Therefore, they recommended to retain the 470–694 MHz band exclusively for broadcasting services (Standeford, 2014). Thus, while the potential mobile allocation of that particular band does not necessarily imply using the band for mobile services as there is already another primary allocation of broadcasting services, allocating the band to mobile services without having all countries agreeing could lead to interference.

11.3 Spectrum Sharing

Spectrum sharing has been part of the RR framework since the early days of wireless communications. For instance, only 0.8% of the European common allocations table consists of exclusive allocations for one radiocommunication service (Forge et al., 2012). More specifically, the RR table of frequency allocation allows each frequency band to be shared by a maximum number of services perceived to be compatible through coordination (ITU-R, 2018). However, the reality is that not all radiocommunication services can coexist with one another (e.g. broadcasting, mobile).

This may be related to the confusion between the application of the RR when it comes to cross-border interference with neighbouring countries and the application of the RR within the same country. This issue was raised prior to WRC-15 with respect to the differences between the RR service allocation and actual national use of radio spectrum (Sweden, 2014). It was eventually clarified that the RR accommodate several frequency bands that are allocated to more than one radiocommunication service even if they are not compatible. Thus, the result of the sharing between existing services allocations and potential additional services allocations should be based on the possibility for cross-border interference and not on the compatibility within a given country. In other words, the RR should not be applied to services that do not cause harmful interference to the stations of another country (Jones, 1968). This was suggested by Coase (Negus & Petrick, 2009), who called for flexibility in spectrum use more than sixty years ago and who argued that flexibility is already accommodated in the RR on the condition of not causing interference to neighbouring countries (Coase, 1959).

The dilemma is that this is not the dominant paradigm within the ITU-R countries. As a consequence, additional service allocation could be restricted due to concerns over the incompatibility with existing services operating in the same band. These concerns may be related to the immaturity of radio technologies during the early day of wireless communications (Horvitz, 2013). Unfortunately, this has led to the focus on protecting against interference that is argued to cause delays to the development of more robust, interference-resistant equipment for many decades (Forge et al., 2012). These deter, in general, spectrum sharing. This is despite the RR not restricting sharing or preventing national regulators from introducing opportunistic access to spectrum, and technologies such as CRS that could operate under the current service allocation (Anker, 2010).

To this end, we suggest several measures to enable greater sharing of radio spectrum. The first of these measures is to have a dedicated band for opportunistic access devices, as operating these devices in spectrum occupied by dominant services, such as broadcasting, is quite difficult. Therefore, it is suggested to have this dedicated band for these devices so that there would be no resistance from the current set of primary users. The idea of a dedicated band was proposed by Lehr and Crowcroft (2005), who argued that dedicated unlicensed spectrum, where all unlicensed devices are considered primary users, will obtain the full benefits of spectrum commons rather than overlay or underlay. It was further argued that underlay or overlay regime are not likely to gain the full benefits of the unlicensed model, as limitations on transmitted power for underlay devices may deter their applicability. This was also supported by Tonge and Vries (2007) who pointed out that the combination of both licensed and licence-exempt allocations will result in a greater collective benefits. Lehr (2004) argued that it is important to allocate dedicated spectrum for unlicensed use rather than focusing only on allowing spectrum easements as the regulatory diversity enable future proof policy.

A second measure could be to allocate dedicated spectrum to experimental services. In fact, this was a concept embedded in the early versions of the RR where there were dedicated bands for experiments (ITU-R, 1927). For instance, in the RR of 1938, several frequencies were used for research and experiments in North America in the 30 MHz to 300 MHz band (ITU-R, 1938). Currently, Article 27 of the RR regulates the operations of experimental stations. More specifically, Article 27.5 clarifies that the rules of the RR shall apply to experimental stations but Article 27.7 provides some flexibility and states that '*[w]here there is no risk of an experimental station causing harmful interference to a service of another country, the administration concerned may, if considered desirable, adopt different provisions from those contained in this Article*' (ITU-R, 2020). In the RR of 2001, the 275–1000 GHz frequency band was labelled for potential use for experimentation with and development of various active and passive services (ITU-R, 2001). It is worth highlighting that the idea of experiential stations has been adopted by several national regulators. For instance, the FCC has a license category for experimental radio service (ERS) to be utilized for experimentation, market trials and product development (Bustamante et al., 2020).

The third measure is to allocate more bands for ISM applications where devices could operate in a similar way to Wi-Fi operating in the ISM bands at 2.4 GHz. The issue here, however, is that the last time the ITU-R decided to allocate bands for ISM was at the WRCs of 1947 and 1959 and, more recently, WRC-79, which accepted an increase in the number of bands to be designated for ISM equipment (ITU-R, 1979). One may ask why sharing and opportunistic access would be more enabled in ISM bands? The answer is that the deployment of spectrum easements approach is, in practice, challenging in any other bands where there are primary services mainly because of the non-fairness of these primary users who tend to limit secondary access to their spectrum. This occurred in the case of TVWS where broadcasters in many countries opposed its introduction into spectrum bands where they operated. Meanwhile, the ISM application did not have the aggressive industry structure that restricted access to ISM bands. If that were the case, we would not enjoy the Wi-Fi services that are available today.

The fourth measure is to enable opportunistic access in a more explicit way in the RR with more rights in higher bands where there is less use of the spectrum while still being secondary to primary services. Having said that, there is a tendency within the ITU-R to allocate more bands for technologies that require exclusive allocation such as IMT and, with the movements towards 5G, the previous WRC-19 discussed several potential spectrum bands in the higher frequencies to be identified for IMT. The danger of such approach is that it prejudges the future of spectrum management and focuses on the exclusive allocation to the mobile telecommunications industry, which goes against the concept of opportunistic access. We propose enabling this measure in higher spectrum band where there is a lot of global allocations. The ITU-R should encourage such an operation by providing it with minimum legal or regulatory status according to the RR so that it has a degree of certainty. This is important, as the current paradigm for many ITU-R countries is that opportunistic access via Article 4.4 is considered to be the opposite of the RR. In other words, there is a perception that the ITU-R focus is on primary services and sharing among countries unlike, for example, the case of TVWS.

The fifth suggested remedy is to map the most common allocation in a specific band onto other bands that have similar characteristics. More specifically, sharing between the different services in the same band is already accommodated in several bands where there are allocations to more than one service. It should be noted that while sharing could be possible in one particular band in one region, this does not mean it is the case for the other ITU-R regions. However, many differences exist due to the historical separations between the ITU-R three regions. This is demonstrated by the observation that in the higher frequency bands there is usually a common service allocation globally.

One application of such a remedy is to apply the coprimary allocation between the mobile and broadcasting service in the 700 MHz band to the rest of the UHF band (470–698 MHz). It is also suggested to apply the most dominant allocation in any of the three ITU regions in the other two regions. Such a step would enable more coprimary allocation across the whole radio spectrum, and, accordingly, provide more flexibility for future radio systems.

Another restriction against having more coprimary or cosecondary services in one band are the limitations that emerge from countries that are geographically separated from other countries within the same region (e.g. Russia is far away from most of the African countries that are also within Region 1) or from countries in neighbouring regions (e.g. Iran lies in Region 3 but has many neighbours in Region 1). The remedy for this could be to apply population distribution maps instead of the current separations between the three regions that was based mainly on the historical disagreement between the USA and Europe (Mazar, 2009).

The sixth suggested measure is to set general service-neutral values for interference parameters across all frequencies at the geographic boundary of every country by way of a multilateral agreement where the ITU acts as an arbitrator in the event that dispute resolution is necessary (Lie, 2004). Another method is to define generic technologies and service models based on anticipated use and reasonable expected receiver performance (Louis, 2011). We understand that this particular remedy is difficult to be applied in practice, but it is suggested here that an appropriate starting point is initially a few frequency bands that could then be expanded if successful.

11.4 Radio Regulations Deficiencies

As with any regulatory framework, the RR are not perfect and accommodate several deficiencies. The main one, in our opinion, is the lack of enforcement procedures to resolve actual cases of interference as the enforcement of the RR is largely driven by its international legal status that countries sign and agree to follow. This is not to say that the ITU-R lacks enforcements measures; they do exist, such as cancelling satellite networks when they violate the RR. However, in cases of intentional interference (jamming) or the reallocation of satellite into orbits that has not been approved, the ITU-R has no enforcement powers (Jakhu & Pelton, 2017).

Therefore, it is argued that by strengthening the role and functions of the ITU-R in terms of how it handles harmful interference, this may provide the confidence necessary for regulators and operators to introduce innovative ways to access spectrum in ways that are not as conservative as has hitherto been the case. More specifically, the way the ITU-R is currently involved in interference resolution should be debated with one suggestion being that it should have a more enforcing role. This could be achieved via establishing regional monitoring stations to determine sources of harmful interference and identify the responsible country(s). However, we must admit that having a greater enforcement role in the RR and the ITU-R is limited by its place within the UN system, where countries often prefer not to empower agencies at the expense of their national sovereignty.

Resolution disputes in the ITU are dealt with by Article 56 of the ITU constitution that suggests either negotiation, diplomatic channels or operating according to procedures established by treaties between countries (ITU, 2019a). The intervention of the ITU-R BR in cases of harmful interference is conditioned by the request of one or more of the interested countries and is limited to conducting investigations and making recommendation to the relevant Member States (ITU, 2019b). There are other procedures for arbitration among Member States if previous measures were not successful. However, they have never been used (Jakhu, 2011). The RRB can also provide non-binding recommendations in cases of harmful interference disputes where unresolved ones are referred to the next WRC where decisions are usually taken based on political considerations rather than RR procedures (Jakhu, 2011).

Secondly, similar to the command and control approach, which has traditionally placed the burden on transmitters to resolve interference, even if receivers are of low performance (Vries & Sieh, 2012), the RR focus is usually on the transmitters. In particular, it is stated in RR Article 15 that transmitting stations shall emit only as much power as is necessary to ensure a satisfactory service (ITU-R, 2008b). On the other hand, the RR state in Article 3.13 that '*the performance characteristics of receivers should be adequate to ensure that they do not suffer from interference due to transmitters situated at a reasonable distance and which operate in accordance with these Regulations*' (ITU-R, 2008a). However, there are no quantified details of either the transmitter power or the characteristics of receivers. Therefore, the RR have traditionally placed the burden on transmitters to resolve interference even if receivers are of low performance so that it cannot reject a signal transmitted on an adjacent channel (Vries & Sieh, 2012).

Other provisions related to the technical performance of frequency assignments nationally can be found within Article 4. Most importantly, Article 4.1 urges Member States to '*limit the number of frequencies and the spectrum used to the minimum essential to provide in a satisfactory manner the necessary services. To that end they shall endeavour to apply the latest technical advances as soon as possible*'. In other words, such provisions encourage maximizing spectrum utilization efficiency and following the latest advancement in wireless communications. The challenge, of course, is that there are no defined measures to assess the applicability of such provisions.

Werbach (2004) argues that it is wrong to put the burden of resolving interference on the transmitter as the problems could also be solved by a more robust receiver. The choice should be whether to have a transmitter that cause less interference or a receiver than can extract the information from the signal based on the welfare effects of both options. Werbach's view is inspired by Coase's contribution 'The Problem of Social Cost'. Coase (1960) assumes the case of a confectioner and a doctor working adjacent to one another. The doctor's work is sensitive to vibrations and is harmed by the confectioner's machinery. Coase explains that it is not only the doctor who is affected by the confectioner but also the confectioner is affected by the doctor's excessive sensitivity. Similarly, interference should not be eliminated; it instead should optimized in a way that the resulting marginal value from it exceeds its marginal cost (Werbach, 2004).

Thirdly, the way interference is managed and perceived is one of the main deficiencies of the RR where the focus is on avoidance and elimination of harmful interference rather than optimizing spectrum usage (Ryan, 2005). In fact, the RR include in Article 1, in addition to the definition of 'harmful interference', definitions of 'permissible interference' when it states that *'[o]bserved or predicted interference which complies with quantitative interference and sharing criteria contained in these Regulations or in ITU-R Recommendations or in special agreements as provided for in these Regulations, and 'accepted interference' Interference at a higher level than that defined as permissible interference and which has been agreed upon between two or more administrations without prejudice to other administrations*'. In our opinion, the focus should be on accepted rather than harmful interference based on the actual amount that occurs in practice.

In fact, practical cases for mitigating interference such as potential interference into terrestrial TV reception when 4G was introduced into the 800 MHz band showed that the interference experienced was only about 1/100th of that predicted (Layton, 2020). This is not to say that interference should be overlocked or underestimated but rather to encourage new measures to manage interference while not assuming the worst case that may not occur. More specifically, fear of interference was related to the inefficiency of wireless devices when wireless telecommunications were emerging. It also aimed to reduce the cost of the mass production of receivers and to encourage investment given the risk of interference and escalating transmitter powers (Horvits, 2012). However, while the situation has changed today, measures of protection against interreference are almost the same as they were a century ago.

Another paradox of the RR is that while there is a recommendation within the RR to allocate spectrum to broadly defined services, it states in other parts that using a broadly defined service where its subset radio services have widely different protection ratios may result in a degradation of performance requirements (ITU-R, 2012c). Therefore, the RR do not implicitly encourage having this broadly defined set of services that allow for greater flexibility in spectrum use.

The main challenge with the RR as it currently stands is that it is a treaty document among ITU Member States, and, therefore, the system is considered a closed one with a lack of transparency (Ryan, 2012). Furthermore, the main spectrum policy principles are supported by elements of the RR which are incorporated by treaty. In addition, the ex ante allocation embedded within the RR could be considered as not being future proof in cases where the emergence of breakthrough wireless technologies or the decline of already approved services or systems which may create artificial scarcity (Ryan, 2012). The ITU has always faced two contradicting challenges: to meet increasingly technological needs of developed countries and to meet the needs of developing countries (Solomon, 1984).

11.5 The Path Towards Change

One key concern is regarding the procedure of introducing changes to the RR and whether it is an easy process. In theory, WRCs are in charge of modifying the RR through pre-agreed agenda items which usually address topical issues related to specific services (e.g. IMT in mobile service, additional frequencies for maritime service) rather than conceptual issues related to the main principles or pillars of the RR. For example, A.I. 1.2 of WRC-12 sought to enhance the international spectrum regulatory framework—see Chapter Four. Such an agenda item was motivated by the convergence of services and the promise of triple play (audio, video and data) enabled by IP.

These extraordinary agenda items have a wide scope and extensive studies without tangible outputs for several reasons. The first is that these discussions are usually led by telecommunications regulators participating in the ITU-R on behalf of their countries, while generally speaking, regulators mostly tend to be inactive unless faced by a real threat to or disturbance in the telecommunications market. Of course, this is not the case for regulators such as Ofcom or FCC which usually encourage new regulatory measures and have the capacity and resources to experiment with them. This was not the case for the RR and the international spectrum management where regulators are in their comfort zone for several decades, and there has not been wireless technologies that call for introducing radical changes. For instance, the ITU-R community discussed CRS and SRD at the time of WRC-12, and there was hype about these technologies but based on the discussions it was decided that there is no need to introduce any changes to the RR.

Secondly, the RR are like a large structure composed of small blocks that were built throughout the last century and it is difficult to introduce any changes into such a structure without severe repercussions. This is in addition to the unknown implications that may have an impact in the future on existing wireless deployments. For instance, the idea of having a converged radiocommunication service was discussed previously in the RR but, due to the potential impact on existing services and uncertainty in applying these changes in practice, such an idea was discarded. Therefore, having additional spectrum allocations for any radiocommunication services where demand is increasing (e.g. mobile service for IMT systems) should take into account the existing use and should not place any extra burdens on the services to which the band is currently allocated to regardless of when this occurred. This is one of the reasons why an additional allocation is usually a complex process.

Moreover, the three geographical regions of the ITU-R were established for historical reasons but they do not accurately reflect the geographical situation where countries in Region 1 are quite close to countries in Region 3 (e.g. Iran, Pakistan). Such geographical separation has caused several problems such as the disagreement over the additional mobile allocation in the 700 MHz band. More specifically, Iran, a Region 3 country, as part of the GE-06 plan that covers Region 1 in addition to Iran, was interested in protecting its terrestrial broadcasting service in the 700 MHz band from the deployment of mobile services in neighbouring Region 1 countries. It could have been more convenient to have Iran as part of Region 1 instead of maintaining the historical division between Regions 1 and 3 that is not geographically based. A closer look at the concept of the three regions also shows that such a separation may not be applicable in the era of globalization but was suitable when the wireless policies of Europe and the USA needed to be separate. Today countries, which disagree on some issues, cooperate in developing wireless standards through their companies and standardization organizations. The 3GPP 5G NR standard is a clear example where Chinese, US and European companies cooperate and have a mutual interest in reaching an agreement.

Thirdly, in the attempts to review the RR, the result has been always that the main principles of these regulations remain valid and are able to meet the challenges emanating from technological advances. For instance, the VGE established to simplify the RR called for retaining the main principles of the RR including maintaining a table of service allocation covering the useable radio-frequency spectrum (ICAO, 1995). Such principles are related to the nature of wireless technologies that are still based on having parts of the radio spectrum allocated to specific services, and, therefore, having a table of service allocations for international radio spectrum that was initially established almost a century ago remains valid.

Fourthly, the dilemma in those articles that provide flexibility to abandon the RR such as Article 4.4 is that it comes at a quite an expensive price, namely accepting interference and not causing interference to systems that may be outdated and inefficient in terms of spectrum utilization. Therefore, these articles may not be practical for several operators and discourage innovation in systems that do not align with the RR. In other words, the RR do discourage wireless technologies that contradict with principles such as radiocommunication service allocation, interference protection and operating in specific bands.

Fifthly, any change in the RR would, quite naturally, be faced with resistance from the incumbent users of the radio spectrum including global satellite operators and mobile operators who have mastered in recent years the regulatory procedures and decision-making process of ITU-R. This is in contrast to those other companies that are not in favour of exclusivity but are not well experienced in the work of ITU-R study groups and WRCs (e.g. Facebook and Microsoft). Therefore, it is argued that the current institutional framework of spectrum management resists any decentralized form of spectrum access because it accommodates organizations that have already expended time and resources to optimize their behaviour according to the current framework (Benkler, 1998). These organizations resist any changes that would entail them incurring new learning costs.

Moreover, it would be difficult for the regulator to recover the spectrum under the current framework without incurring costly compensation to operators and end users (Werbach, 2004). In addition, highly fragmented usage rights may cause delays to the aggregation of spectrum into other usages (FCC, 2010). Herzel (1951) considers many reasons that make any reform of the regulations in the market difficult, for example, the strangeness of new ideas, especially those associated with less government control and lobbying by groups that have interest in the current method of regulations. It is argued that existing institutional framework and arrangements are based on significant learning effects where any change to them will require time and resources to optimize their behaviour (Benkler, 1998).

Marcus (2004) argues that historically governments have suppressed more technologies than they have promoted. For example, although spread spectrum technology was invented in the 1940s, it was classified by the government until it was 'reinvented' in the 1960s. Similarly, Heller (2008) explains that although cellular mobile technology was initially developed in 1947, the FCC started making rules for the service in 1968 and began issuing licenses in 1982. Such regulatory gridlock is considered to be a barrier to innovation in spectrum use (Lessig, 2001).

It is also argued that the RR have become a tool for the main stakeholders internationally to block spectrum uses that are not in their interest (Kelly, 2002). For instance, there have been several attempts to modify the current RR structure and functions during WRC-1995, WRC-03, WRC-07, WRC-12 and WRC-15, and each time there was a perception that there is a need to modify the RR to accommodate greater flexibility and to handle the deficiencies in the international service allocation framework. However, each time it was agreed by most of the ITU-R countries to retain the current situation with effectively no changes being made. With respect to industry, the significant investment has been made in the existing radio system may postpone the mass production of any other radio device even if it is more spectrum efficient (Struzak, 2007a).

Furthermore, one perception of the current spectrum management framework is based on the historical model of wireless broadcasting which is centred on powerful transmitters on the operator side and simple receivers on the end user's side (Benkler, 1998). Another perception is that such a framework is related to the radio technology that Marconi developed, which required high signal-to-noise ratios, and, accordingly, a high degree of protection against interference by assigning exclusive privileges (Lie, 2004). Having mentioned said, it is not a surprise that WRCs usually favour the protection of existing systems and applications (Radiocommunication Bureau, 2007). It is reasonable to expect that existing service providers would lobby to resist any additional allocation on a coprimary or cosecondary basis. This was quite apparent in WRC-15 where the broadcasting industry opposed any coprimary allocation in the UHF band (World Radiocommunication Report, 2015a).

Another reason for the resistance is related to the perception of new technologies for opportunistic access and spectrum easements. There were high expectations, but not unrealistic ones, with regard to CRS (Forde & Doyle, 2013). In fact, this was a key problem, especially for regulators who embraced a type of management that enabled strict control of spectrum as it limited their willingness to move towards technologies that were perceived to access any part of the spectrum without the need for approval by regulators.

In addition, calls for radical changes in spectrum management are not currently supported by practical technologies that can radically alter the way we use spectrum and, accordingly, the way we manage spectrum. In other words, researchers calling for open access to spectrum (e.g. Benkler, 1998; Noam, 1995) are influenced by the Internet model but, at the end of the day, spectrum use and management are restricted by the laws of physics (Struzak, 2007a). On the one hand, the development of wireless technologies is influenced by the traditional model of transmitting and receiving, but on the other hand, advocates for change are usually policy makers or scholars influenced by other model(s) of managing resources that are different in nature than the radio spectrum.

We believe that introducing changes and reforming the RR is possible under specific circumstances. The first of them is to have an influential or entrepreneurial country (or groups of countries) to promote one of these changes or reform measures in a future WRC. This should occur after introducing these changes nationally given that while national changes of frequency planning could take up to two years, international changes can take up to 10 years (Berggren et al., 2004). This is what happened with Wi-Fi in the ISM bands where the transformation was led by the USA. More specifically, while the ISM bands were originally allocated in WRC 1947 (ITU-R, 1947), that decision seems to be taken shortly after the USA adopted a similar decision (USA, 1948). Moreover, this band was initially allocated to ISM in the RR of 1947 in Region 2, where the USA is located, and in some countries in Regions 1 and 3. It was then allocated globally in the RR of 1982 (Ard-Paru, 2012). A major change in the use of these ISM bands was when the FCC decided in 1985 to open up three ISM spectrum bands (915 MHz, 2.4 GHz and 5.8 GHz) for wireless local area network (WLAN) (Lemstra et al., 2011a). The decision led to a similar transformation in 2.4 GHz in Europe (Lemstra et al., 2011b).

Another example of policy transformation was in 1992 when the 5.150–5.350 GHz spectrum band was allocated in Europe to WLAN (CEPT, 1992). The industry in the USA was emboldened by this decision and called in 1995 for a similar allocation in the 5 GHz band for WLAN. Consequently, the FCC allocated the 5150 to 5350 MHz and 5725 to 5825 MHz spectrum bands for the operation of the unlicensed national information infrastructure (U-NII) devices (Anker & Lemstra, 2011). Following this, WRC-03 decided to allocate the band on a primary basis for mobile services (ITU-R, 2003). In such a case, the Wi-Fi regulations in the 5 GHz involved adopting the European regulations on what they called HIPERLAN and turning them into international regulations mentioned in the RR (CEPT, 1992).

In both cases, the transformation of a particular concept was first initiated by an entrepreneurial country with their domestic preferences then being adopted by other countries. The international platform of the RR was then used to promote such a concept internationally. In fact, the transformation of the international telecommunication regime has also been through similar steps (Ratto-Nielsen, 2006). In the case of transforming monopolies into competitive markets, the international telecommunication regime influenced national policy preferences and also reflected the domestic preferences of the majority of Member States. Finnemore and Sikkink (1998) emphasize the role of entrepreneurs and the existence of international organizational platforms to promote new concepts. If a critical mass of countries supports new concepts, there could be a transformation internationally as suggested by the 'loop effect' of Ratto-Nielsen (2006).

Policy diffusion and adoption in spectrum management is possible between developed and developing countries to a certain extent based on state characteristics, government involvement in the telecommunications market, and its links with these developed countries or regional groups (Wavre, 2018). A leader and entrepreneurial country calling for change within the RR could be the USA, as it lies in a different region (Region 2) than the other main stakeholders in the ITU-R (Europe in Region 1 and Asia in Region 3), which could be useful for any possible transformation within the RR. In particular, this would enable the USA to be sufficiently autonomous in order to experiment with new spectrum use models and technologies. In fact, the USA was the first Member State to challenge the main principles of the international telecommunication regime considering that it was not one of the founding members and it did not have the same regulatory structure as other Member States, and it was one of the first to allow private ownership of telecommunication networks (Ratto-Nielsen, 2006).

For such changes to be transferred to other countries and prevail around the globe, it is important to include them within the RR to obtain acceptance for these changes. Evidence of the influential effect of the RR can be found in the fact that there has only been a few violations of the RR (Zacher, 1996). This extends to the ITU-R recommendations, which, although they are not treaty instruments, are still widely adopted and most countries comply with them most of the time (Ratto-Nielsen, 2006). Having said that, it is important to include the ITU-R and accommodate within the RR resolutions calling for such change or reform of international spectrum management. This was mainly the case in the telecommunication policy reform movements in 1990s where international organizations played a critical role in the diffusion of liberalization and privatization policies (Rodine-Hardy, 2013). This was achieved by influencing preferences for reform, providing forums for policy discussions, and by providing forums for the emergence of new policy standards.

Another caveat when reforming the RR is such change, even if it is led by an entrepreneurial country or group of countries, should preferably be driven by a technology rather than idea or concept. The main traditional principles that the RR are currently based on relate to inefficiencies within wireless technologies in terms of interference immunity. Therefore, as long as radio devices have a limited tuning range, imperfect interference tolerance, and inadequate receiver characteristics, we should not expect any major change to the main principles of the RR such as service allocation. That was evident in the unliteral approach of the USA when deploying IMT in the 28 GHz and how it was not adopted by other countries considering that it was not included in the RR. The US divergence from the RR was not embedded in devices or technologies that are capable of overcoming the difficulties of sharing or compatibility with other systems. Instead, it is dependent on US demands and spectrum usage. In other instances, the USA, supported by its telecommunications industry, was quite keen to have global support for its national policies in order to create a bigger market.

This was quite apparent in the case of Wi-Fi where devices are smart and have different channel plans in the 2.4 GHz band according to country and regional regulations. The USA was also successful in the case of WLAN in 5 GHz at the time of WRC-03 where sharing measures in the band with systems such as radar were incorporated into the RR. In both cases, such an individual approach was embodied in wireless technologies that several countries imported, and whose citizens use both domestically and internationally. Another example of how technology can overcome boundaries and enforce a particular spectrum use is in Russia following the fall of communism where the illegal use of imported two-way radio interfered with military systems (Horvitz, 2009).

One may wonder if such change is likely to happen in the near future? Unfortunately, it seems that change is not perceived as beneficial for the current radio spectrum stakeholders who appear quite keen to retain the current interference management model based on exclusivity and average devices characteristics in terms of receivers and tuning range. Although some countries such as the USA are advancing spectrum policies that encourage sharing (e.g. CBRS), traditional spectrum stakeholders are not positively responding. A way out of this dilemma could be facilitated by non-traditional stakeholders such as Amazon and Facebook, who it appears are interested in a commons approach to radio spectrum. However, those stakeholders need to consider the political economy of decision-making in spectrum policy where service operators who call for exclusivity to internalize the full benefit of their investment (Benkler, 2012).

Having said that, revisiting the history of spectrum management it seems that crisis are one of the main ways to introduce changes to the way we manage spectrum internationally. For instance, it was the Titanic tragedy that led to the Berlin Treaty of 1912 requiring radio equipment and operators on passenger steamers and spurred the US Congress to pass the Radio Act of 1912 that required the licensing of operators and stations (The IEEE—EMC Society, 2007), and it was the chaos of the broadcasting industry in the USA that stimulated the passing of the Radio Act of 1927 (Marcus, 2004). More recently, the tragedy of the Malaysian Airlines flight MH370 in changes within the RR to support global flight tracking. As Milton Friedman explained '*[o]nly a crisis—actual or perceived—produces real change. When that crisis occurs, the actions that are taken depend on the ideas that are lying around*' (Grayson, 2020).

One last remark is that the system of the ITU-R has its own internal tool to introduce changes when needed if requested and approved by Member States. One valid suggestion is to empower the ITU-R to become the international telecommunication regulatory authority (Jakhu, 2000), but it is not expected that only Member States would request or approve this in the light of increasingly weakness of the role of international organizations and the withdraw of the USA from important organizations such as the UNESCO and WHO. Even though the RR are a treaty document, any Member State can file a 'reservation' stating that they will not be bound by a particular regulation (Horvitz, 2009). In other words, the Member States of the ITU-R have the tools to both empower the organization and also to limit its role.

To this end, the ITU should develop links with other organizations involved in telecommunications policy including spectrum aspects (e.g. World Bank, Open Spectrum). In addition, the three ITU sectors should harmonize more their activities not to find a situation where unlicensed and open access to spectrum heavily participate in the ITU-D while the 'old guard' (e.g. incumbent mobile manufacturers and satellite operators) have their reserved seats in the ITU-R. There should be a common unit in the ITU that holistically addresses technology, applications and spectrum.

The international RR should also reflect of our future needs. They are, in some cases, a snapshot of our mistaken predictions of the future. Therefore, there is a need for a continuous review of the main service allocations within these regulations, and a periodic examination of the main principles that underpin them. In an era where telecommunications products and services are widespread, there is a need to have flexible frameworks rather than detailed regulations to cope with advancements of wireless technologies and services. The fact that these regulations are based on the same principles adopted more than a century ago necessitates calls for revisiting these principles and examining whether they deter innovation in the way we utilize radio spectrum. Spectrum utilization is related to the way we manage spectrum which is, in turn, greatly influenced by the principles embedded within the RR. While there have been several reviews of the management and administration of the ITU without addressing the RR principles (e.g. Duque-Gonzalez & Othman, 2001; Tarasov & Achamkulangare, 2016), there is a clear need to go one step further.

At the beginning of this book, we highlighted James Savage's quote that the ITU does not receive much attention. This is exactly the issue that we have sought to address with this book. The RR deserve more attention from researchers and commentators as they have a great influence beyond being simply a treaty document on the way we develop wireless technologies and shape access to spectrum. The latest radical change to the ITU was suggested by the High-Level Committee (HLC) that was appointed in November 1989, and which suggested several structural measures (e.g. elected Bureau director, replacing IFRB, abolishing CCIR) (Solomon, 1991). The HLC recommended that the Radiocommunications Conference should review and revise the RR as necessary without needing to revise their main principles (Savage, 1991). Holding a special WRC to design a long-term plan for the different terrestrial and space services in a more conceptual way is an attractive and practical suggestion (Jakhu & Pelton, 2017). Such a WRC would not focus on specific items as these are usually addressed through agenda items but it would rather concentrate on the principles of the RR and involve academia and NGOs as well as ITU Member States to discuss those innovative technologies and solutions that are typically omitted at WRC. Through doing so, it would seek to ensure the continued relevance of the RR.

References

Anker, P. (2010). *Cognitive radio, the market and the regulator*. Paper presented at the IEEE Dyspan 2010 conference, Singapore.

Anker, P., & Lemstra, W. (2011). Governance of radio spectrum: License exempt devices. In W. Lemstra, V. Hayes, & J. Groenewegen (Eds.), *The innovation journey of Wi-Fi: The road to global success*. Cambridge University Press.

Ard-Paru, N. (2012). *Information and coordination in international spectrum policy: Implications for Thailand*. Retrieved from https://research.chalmers.se

Benkler, Y. (1998). Overcoming agoraphobia: Building the commons of the digitally networked environment. *Harvard Journal of Law and Technology, 11*(2), 1–113.

Benkler, Y. (2012). Open wireless vs. licensed spectrum: Evidence from market adoption. *Harvard Journal of Law and Technology, 26*(1), 69–163.
Berggren, F., Queseth, O., Zander, J., Asp, B. R., Jönsson, C., Stenumgaard, P., Kviselius, N. Z., & Wessel, J. (2004). *Dynamic spectrum access: Phase 1: Scenarios and research challenges.* Retrieved from http://thorngren.nu
Bustamante, P., Weiss, M., Sicker, D., & Gomez, M. M. (2020). Federal Communications Commission's experimental radio service as a vehicle for dynamic spectrum access: An analysis of 10 years of experimental licenses data). *Data & Policy, 2*(6), 1–13.
CEPT. (1992). *Recommendation T/R 22–06. Harmonised radio frequency bands for high performance radio local area networks (HIPERLANs) in the 5 GHz and 17 GHz frequency range,* https://www.etsi.org
Coase, R. H. (1959). The federal communications commission. *Journal of Law & Economics, 2*(1), 1–40.
Coase, R. H. (1960). The problem of social cost. *Journal of Law & Economics, 3*(October), 1–44.
Cronk, M. (2011). Spectrum regulation and bandwidth. *EBU technical review* (October).
Duque-Gonzalez, A., & Othman, K. (2001). *Review of management and administration in the International Telecommunication Union (ITU).* Retrieved from https://www.unjiu.org
FCC. (2010). *Connecting America: The national broadband plan.* Retrieved from https://www.fcc.gov
Finnemore, M., & Sikkink, K. (1998). International norm dynamics and political change. *International Organization, 52*(4), 887–917.
Forde, T., & Doyle, L. (2013). A TV whitespace ecosystem for licensed cognitive radio. *Telecommunications Policy, 37*(2–3), 130–139.
Forge, S., Horvitz, R., & Blackman, C. (2012). *Perspectives on the value of shared spectrum access: final report for the European commission.* Retrieved from http://ec.europa.eu
Grayson, D. (2020). Only a crisis—actual or perceived—produces real change, *Thinkunthink.* Retrieved from https://www.thinkunthink.org/
Heller, M. (2008). *The gridlock economy: How too much ownership wrecks markets, stops innovation, and costs lives.* Basic Books.
Herzel, L. (1951). 'Public interest' and the market in color television regulation. *University of Chicago Law Review, 18*, 802–16.
Horvitz, R. (2009). *Towards an open ITU.* Retrieved from https://www.researchgate.net/
Horvits, R. (2012). *Perspectives on the value of shared spectrum access.* Paper presented at the 3rd PolicyTracker Middle East spectrum conference, Manama.
Horvitz, R. (2013). Geo-database management of white space vs. open spectrum. In E. Pietrosemoli, & M. Zennaro (Eds.), *Tv white space: A pragmatic approach.* ICTP.
ICAO. (1995). *ICAO position for the ITU.* Paper presented at the world administrative radiocommunication conference (WARC-95), Geneva.
Indepen, A. (2001). *Implications of international regulation and technical considerations on market mechanisms in spectrum management: Report to the independent spectrum review.* Retrieved from http://www.ofcom.org.uk
ITU. (2019a). *Constitution of the international telecommunication union.* ITU, Geneva. Retrieved from www.itu.int
ITU. (2019b). *Convension of the international telecommunication union.* ITU, Geneva. Retrieved from www.itu.int
ITU-R. (1927). Article 5: Radiotelegraph regulations. In *General regulations annexed to the international radiotelegraph convention.* ITU.
ITU-R. (1938). *General radiocommunication regulations.* Paper presented at the international telecommunication convention, Cairo.
ITU-R. (1947). *General radiocommunication regulations* Paper presented at the international telecommunication convention, Atlantic City.
ITU-R. (1979). WRC-79 resolution 63. Relating to the protection of radiocommunication services against interference caused by radiation from industrial, scientific and medical (ISM) equipment. In *Final acts—world radiocommunication conference (WRC-79).* ITU, Geneva.
ITU-R. (2001). Article 5: frequency allocations. In *WRC-2000 final acts.* ITU, Geneva.

ITU-R. (2003). Resolution 229: Use of the bands 5 150-5250 MHz, 5250-5350 MHz and 5470-5725 MHz by the mobile service for the implementation of wireless access systems including radio local area networks. In *Provisional final acts—world radiocommunication conference (WRC-03).*
ITU-R. (2008a). Article 3: Technical characteristics of stations. In *Radio regulations* (Vol. 1). ITU, Geneva.
ITU-R. (2008b). Article 15: Interferences. In *Radio regulations* (Vol. 1). ITU, Geneva.
ITU-R. (2012a). Appendix 10: Report of harmful interference. In *Radio regulations* (Vol. 1). ITU, Geneva.
ITU-R. (2012b). Article 9: Procedure for effecting coordination with or obtaining agreement of other administrations. In *Radio regulations* (Vol. 1). ITU, Geneva.
ITU-R. (2012c). WRC-12 Recommendation 34. Principles for the allocation of frequency bands. In *Radio regulations* (Vol. 4). ITU, Geneva.
ITU-R. (2018). Report ITU-R SM.2093–3: Guidance on the regulatory framework for national spectrum management. In *ITU-R reports SM-Series*. ITU, Geneva.
ITU-R. (2020). Article 27: Experimental stations. In *RR-2020*. ITU.
Jakhu, R., & Pelton, J. (2017). *Global space governance: An international study*. Springer.
Jakhu, R. S. (2000). *International regulatory aspects of radio spectrum management*. Paper presented at the workshop on 3G reforms: Policy and regulatory implications, India.
Jakhu, R. S. (2011). *Dispute resolution under the ITU agreements*. Retrieved from https://swfound.org
Jones, W. K. (1968). Use and regulation of the radio spectrum: Report on a conference. *Washington University Law Review, 1968*(1), 71–115.
Kelly, T. (2002). Never-ending international telecommunication union reform. In R. Mansell, R. Samarajiva, & A. Mahan (Eds.), *Networking knowledge for information societies: Institutions & intervention*. Delft University Press.
Layton, R. (2020). GPS interference fears are today's Y2K, says former UK spectrum director. *Forbes*.
Lehr, W. (2004). *Dedicated lower-frequency unlicensed spectrum: The economic case for dedicated unlicensed spectrum below 3 GHz*. Retrieved from https://www.newamerica.org
Lehr, W., & Crowcroft, J. (2005). *Managing shared access to a spectrum commons*. Paper presented at the IEEE symposium on new frontiers in dynamic spectrum access networks, Baltimore.
Lemstra, W., Groenewegen, J., & Hayes, V. (2011a). The case and the theoretical framework. In W. Lemstra, V. Hayes, & J. Groenewegen (Eds.), *The innovation journey of Wi-Fi: The road to global success*. Cambridge University Press.
Lemstra, W., Links, C., Hills, A., Hayes, V., Stanley, D., Heijl, A., & Tuch, B. (2011b). Crossing the chasm: the Apple AirPort. In W. Lemstra, V. Hayes, & J. Groenewegen (Eds.), *The innovation journey of Wi-Fi: The road to global success*. Cambridge University Press.
Lessig, L. (2001). *The future of ideas: The fate of the commons in a connected world*. Random House.
Lie, E. (2004). *Radio spectrum management for a converging world*. Paper presented at the workshop on radio spectrum management for a converging world, Geneva.
Louis, J. (2011). *International radio spectrum management beyond service harmonisation*. Paper presented at the Fourth International Conference on Emerging Trends in Engineering & Technology. Port Louis, Mauritius.
Louis, J., & Mallalieu, K. (2007). *Investigating the Impact of convergence on the international spectrum regulatory framework*. Paper presented at the proceedings of the second international conference on systems and networks communications ICSNC '07 Washington D.C.
Marcus, B. K. (2004). The spectrum should be private property: The economics, history, and future of wireless technology. *Essays in political economy, Ludwig von Mises Institute*.
Mazar, H. (2009). *An analysis of regulatory frameworks for wireless communications, societal concerns and risk: The case of radio frequency (RF) allocation and licensing*. (Ph.D.). Middlesex University.

Negus, K. J., & Petrick, A. (2009). History of wireless local area networks (WLANs) in the unlicensed bands. *info, 11*(5), 36–56.
Noam, E. (1995). Taking the next step beyond spectrum auctions—open spectrum access. *IEEE Communications Magazine, 33*(12), 66–73.
Radiocommunication Bureau. (2007). Report of the director on the activities of the radiocommunication sector on resolution 951. In *World radiocommunication conference (WRC-07)*. ITU, Geneva.
Ratto-Nielsen, J. (2006). *The international telecommunications regime: Domestic preferences and regime change*. Retrieved from www.lulu.com
Rodine-Hardy, K. (2013). *Global markets and government regulation in telecommunications*. Cambridge University Press.
RSPG. (2013). *RSPG report on furthering interference management through exchange of regulatory best practices concerning regulation and/or standardisation*, https://rspg-spectrum.eu/
Ryan, P. (2005). The Future of the ITU and its Standard-Setting Functions in Spectrum Management. Standards Edge: Future Generation, Sherrie Bolin, ed., p. 341, Sheridan.
Ryan, P. S. (2012). The ITU and the internet's titanic moment. *Stanford Technology Law Review, 2012*(8), 1–36.
Savage, J. (1991). The high-level committee and the ITU in the 21st century. *Telecommunications Policy, 15*(4), 365–371.
Solomon, J. (1984). The future role of international telecommunications institutions. *Telecommunications Policy, 8*(3), 213–221.
Solomon, J. (1991). The ITU in a time of change. *Telecommunications Policy, 15*(4), 372–376.
Standeford, D. (2014). EU high level group divided on future use of 470–790 MHz. *PolicyTracker*. Retrieved December 1, 2020, from www.policytracker.com
Struzak, R. (2007a). *Flexible spectrum use and laws of physics*. Paper presented at the ITU workshop on market mechanisms for spectrum management, Geneva.
Struzak, R. (2007b). *Spectrum management & regulatory issues*. Paper presented at the ITU workshop on market mechanisms for spectrum management, Trieste.
Sweden. (2014). *General consideration regarding allocation and use*. Paper presented at the meeting of fifth meeting of joint task group 4-5-6-7, Geneva. Retrieved from www.int.itu
Tarasov, G., & Achamkulangare, G. (2016). *Review of management and administration in the International Telecommunication Union (ITU)*. Retrieved from https://www.unjiu.org/
The IEEE – EMC Society. (2007). EMC History Through the Decades. Retrieved November 30, 2014, from http://simbilder.com
Tonge, G., & Vries, P. D. (2007). The role of licence-exemption in spectrum reform. *Communications and Strategies, 67*, 85–106.
USA. (1948). *1947 Supplement to the code of federal regulations of the United States of America. Title 43–Title 50*. United States Government Printing Office.
Vries, P. D., & Sieh, K. A. (2012). Reception-oriented radio rights: Increasing the value of wireless by explicitly defining and delegating radio operating rights. *Telecommunications Policy, 36*(7), 522–530.
Wavre, V. (2018). *Policy diffusion and telecommunications regulation*. Palgrave Macmillan.
Werbach, K. (2004). Supercommons: Toward a unified theory of wireless communication. *Texas Law Review, 82*, 863–973.
World Radiocommunication Report. (2015a). Ignoring administrations supporting flexibility said not a reasonable approach. *World Radiocommunication Report, 6*(43).
World Radiocommunication Report. (2015b). Proponents pressed for flexibility on a country-by-country basis in lower UHF. *World Radiocommunication Report, 6*(43).
Youell, T. (2014). Brussels creates high level group to examine UHF bands. *PolicyTracker*. Retrieved December 1, 2020, from www.policytracker.com
Zacher, M. W. (1996). *Governing global networks: International regimes for transportation and communications*. Cambridge University Press.